Dr. E. Norbert Smith.

Creation or Evolution?
Consider the Evidence Before Deciding

by

E. Norbert Smith, Ph.D.

Printed in the United States of America

Smith, E. Norbert   1941

 ISBN-13: 978-1456468279

ISBN-10: 1456468278

Lord Francis Bacon

Read not to contradict and confute; nor to believe and take for granted; nor to find talk and discourse; but to weigh and consider. Lord Francis Bacon (1561-1626)

Dedication

Henry M. Morris, Ph.D.

This book is dedicated to the fond memory of Dr. Henry M. Morris. He was former president and one of the ten founders and of the Creation Research Society in 1963. This is one of the oldest and most respected creation science organizations in the world. Its publication, the ***Creation Research Society Quarterly***, continues to be the leading scientific journal publishing Creation and Flood related research. In 1970, he founded the Institute for Creation Research and served as its first President. This organization continues to be a powerful force in the origins debate and is perhaps best known for providing expert lecturers supporting Creation science and participating in debates at university campuses around the world.

Dr. Morris was a gentle, soft-spoken man and helped lead the way in modern scientific apologetics. He was a major driving force in starting the Creation science revolution that continues today. The long time best selling classic scientific apologetic book, *The Genesis Flood* written by Henry Morris and Dr. John Whitcomb, Jr. was first published in 1961. It showed there was no need to apologize for the science in the Bible and that the actual scientific evidence supporting macroevolution is nonexistent. That important book had a major impact on my life as it has for thousands of others.

I had the distinct pleasure of serving under his leadership on the board of directors of the Creation Research Society as well as teaching a graduate course at the Institute of Creation Research while he was president. His dedication and life-long commitment to biblical inerrancy touched countless lives around the world and many will be eternally grateful. *He who pursues righteousness and love finds life, prosperity and honor.* (Proverbs 21:21) These words certainly applied to Dr. Henry Morris. He will be missed for a long time, but his legacy lives on.

Table of Contents

Foreword

It is my distinct pleasure to write a brief word for this new book on the evidences for creation. Dr. Smith has compiled a great deal of information on a topic vital for a proper understanding of our own personal life and the health of a nation. Without a correct understanding of our origins, we cannot find answers to life's big questions. Where did I come from? Where am I going? What is the worth of life? What is the meaning of life? Where do I go after I die? Answers to these questions differ widely depending on each person's perspective. Are we merely the mutated, chance by-product of mindless evolutionary processes, or are we created in the image of God, with great worth in his eyes and an important destiny to perform?

In recent years, scores of research scientists **and** university professors have banded together to study this crucial subject of origins from a Christian perspective. These scientists are fully qualified in their fields, but they are not ordinary scientists. They have come to conclusions which give them an advantage over many others. They have seen the evidence for, and placed their trust in, the written Word of God, the Bible. They are not blindly following an unfounded faith; their study has led each to conclude that the Bible contains true history about the world's beginnings, events otherwise outside the realm of scientific observation. If its claims are true, it provides the correct framework within which we must conduct all our thinking, even scientific thinking. It does not give us all the details, but it does identify several major events of history, which affected the entire planet and everything in it. If we ignore these world-altering episodes, we have no chance to accurately reconstruct any fact or process that occurred in the past or occurs in the present.

The most crucial aspect of scientific investigation often goes unnoticed or unstated. One's presuppositions at the start dictate the conclusions. Particularly as it relates to unobserved, unrepeatable events of the past, if we begin with the (often unacknowledged or unstated) faith position that only *natural* processes have ever taken place on Earth, then our conclusions will reflect that thinking. For example, no matter how complex or engineered or designed a life form is seen to be, we must interpret it as the result of unthinking, unguided, *natural* processes (like *natural* selection). One's bias has forbidden him from concluding that the design came from a super*natural* Designer, even if it may be obvious. Science now knows that even a tiny single celled organism is more complex, more engineered, more designed than a super computer! How dare we conclude that things more complex than a super computer are the chance results of blind, natural processes? Yet this is a common view among today's scientists. This is the faith position of *naturalism* taught in our textbooks in the name of science.

However, as you will see in this book, the scientific evidence strongly favors creation. In a technical sense we can prove neither view of the past, for our observations are limited to the present. But the observations we do make in the present agree with the presuppositions of creation thinking far better than with the presuppositions of the evolution model. The evidence fits the creation idea about history far better than the evolution idea about history. When the two competing ideas about history are compared, creation is far superior.

We see then, that both evolution and creation are worldviews. We also see that creation is at least as scientific as evolution, and evolution is at least as religious as creation. We must choose one to be our guide in life, and answer life's big questions for us. My conviction and the conclusion of this book is that creation makes far better sense of both life and science than evolution. It alone provides a sound platform upon which one can build a life and stake eternal destiny.

John Morris, Ph.D.
President, Institute for Creation Research www.icr.org

Acknowledgements

Many people throughout my life have contributed to this effort. My mother, Opal Smith and grandparents Hans and Emma Hansen instilled in me an early love and respect for all things living. My son, Jayson, spent untold hours teaching his computer illiterate father how to better use my old computer and the Internet. Jayson was relentless in impressing me with the value of backing up computer files on disks enabling early drafts of this book to survive two computer crashes and the theft of my home computer. Thanks, Jayson, for your tireless effort and unwavering patience with your sometimes forgetful and often annoying father. My daughter, Weena teaches high school English and has long emphasized the importance of using proper English, instead of my quaint Oklahoma dialect. She helped edit an early version of this book.

In 1964 Paul Ehmer, a co-worker at Texas Instruments, gave me a copy of **The Genesis Flood** by Henry Morris and John Whitcomb. That book forever changed my life. Years later, I had the privilege of serving on the board of directors for the Creation Research Society and worked closely with Henry Morris and John Whitcomb as well as Henry's son John and several other leading creationists including Wayne Frair, George Howe, Emmett Williams, Duane Gish and others. I have stayed in touch with these Creation science leaders over the decades and thank each of you for your inspiration and continuing friendship. We are truly at war and our enemy is unscrupulous. As the Apostle Peter warned us, we must always be on guard. ***Be self-controlled and alert. Your enemy the devil prowls around like a roaring lion looking for someone to devour. Resist him, standing firm in the faith, because you know that your brothers throughout the world are undergoing the same kind of sufferings.*** (1 Pet 5:8-9, NIV)

An impressive cadre of professional acquaintances and friends spent untold hours reviewing early versions of this book and I sincerely thank each of you. It is better because of your effort. Some I have known for decades, others I met recently online. The following people have reviewed all or part of the manuscript and have given permission to mention their names. Due to the controversial nature of the topic, others preferred to remain anonymous. I certainly understand and respect their concern.

My long time botanist friend, George Howe, Ph.D. has been an enduring mainstay in the modern Creation science movement and served as President of the Creation Research Society. Dr. Howe held several important offices in that organization including serving as a member of the board of directors for more than forty years. He served as biology editor for the prestigious ***Creation Research Society Quarterly*** until his recent retirement. He is a respected scientist and has published dozens of scholarly research articles in various scientific journals. I have known Dr. Howe since my days as a doctoral student at UCLA over three decades ago when he was my beloved Sunday school teacher. He made many suggestions for improving this book and has encouraged me over the decades. Thank you Dr. Howe.

Frank Sherwin of the Institute for Creation Research (ICR) made significant contributions to several chapters of this book. He was especially helpful with the history of the creation movement and impact of Dr. Henry Morris. Thank you for your tireless effort, encouragement and friendship over the years. ICR remains an important leader in the current origins debate.

Stan Robertson has been my best friend for over three decades and we have discussed origins countless times, yet still have widely different worldviews. This Bible verse applies to our relationship: ***As iron sharpens iron, so one man sharpens another.*** (Proverbs 27:17) Stan, thank you for your friendship and for the many lively discussions on diverse topics we have had over the years. We have proved many times over that it is possible to disagree, without being disagreeable. We met at Fort Hays State University in Kansas where I taught a biology course for non-science majors after earning my doctorate. Later, we taught together at Northeastern Oklahoma State University. He is a physicist with uncanny math skills and we have published technical papers together dealing with blood flow in alligator skin and related matters. For the past several years he taught physics nearby in Weatherford, Oklahoma and our friendship continued to grow. He recently retired as Emeritus Professor of Physics, Southwestern Oklahoma State University and moved to be closer to his grandchildren. He is sorely missed. Stan thanks for your help with this project and for your enduring friendship over the years.

I also met Jerry Bergman, Ph.D. nearly thirty years ago. He came to a board of directors meeting for the Creation Research Society in Michigan shortly after his dismissal from Bowling Green State University. He was one of the first to suffer religious persecution in the United States for rejecting evolution. The major reason for his dismissal, one of his detractors noted, was for writing a monograph that mentioned Creation and evolution. He was not advocating either view in the monograph published by *Phi Delta Kappa*, the education honor society. Just the mention of Creation cost him his job, but unlike many others, he was not deterred and remained in academia.

Jerry and I have stayed in touch over the years and he is the most productive scientist and author I know. He collects degrees like some children collect stamps. At last count, he had numerous BS degrees, five Master's degrees, two doctorates along with several post doctoral research experiences. He has taught life science courses at the graduate and undergraduate levels and has published over a thousand technical articles, books, and popular magazine articles. He is a recognized leader in the Intelligent Design movement and recently published the first in a six volume series, ***Slaughter of the Dissidents*** about the prejudice and religious persecution targeting anyone doubting Darwin. In preparation for writing this well documented reference, he interviewed over 300 professors and others that have lost their careers for expressing doubts about evolution. He knows hundreds more cases, but they feared repercussions for going public and remain in the closet. That fear is well founded. To date, over 3,000 university and high school teachers have been denied tenure or outright fired for expressing doubts about evolution dogma. Dr. Bergman also has a website and over 3,000 former evolutionists openly reject evolution dogma. (See: www.rae.org/darwinskeptics.html). In spite of what many people believe, religious persecution abounds in the United States today and freedom of speech does not apply regarding one's views of the origin of life. Thanks Dr. Bergman for your helpful comments regarding this book and for your friendship and encouragement over the decades.

Historian Fritz Ward, Ph.D. has two doctorates and like me was a former truck driver. We met at a truck stop while refueling our big rigs next to each other. Due to the excellent pay, there are many educated truck drivers on the road. We stayed in touch and our friendship has deepened over the years. He is also active in the Intelligent Design controversy and currently teaches math to elementary students at a major charter school in California. Thanks for your friendship and inputs regarding this book.

Steve W. Pray, Ph.D., D.Ph. is a church friend. He teaches at Southwestern Oklahoma State University as the prestigious Bernhardt Professor of nonprescription products and devices. He poured over very word of the manuscript and offered many suggestions for improvement. I sincerely appreciate his efforts.

William A. Cirignani, Esq. is a new online friend. He is a Christian attorney practicing law in Chicago, Illinois and spent many hours reviewing the manuscript and exchanging email comments. His non-science background and skills in word use and logic gave him insight into areas others missed. Thank you for your effort, the book is better because of your input.

Brian D. Sherman, D.Ph. and I met some years ago though New Life Church in Colorado Springs and have grown closer. Our hearts beat as one regarding origins and many other areas. He has recently written a screenplay, *Trucker U* based loosely on my tenure denial story and hopes to sell the movie rights for production. He also offered suggestions on ways to improve the readability of this book. Thanks Brian, you are a good friend and kindred spirit.

Several others helped with various aspects of this work and I sincerely appreciate their efforts. Paul Brown and Anna Anderson helped in the review process. My hired grammar gun, Michelle Whitefield deserves special mention. English is one of my areas of accomplished ignorance. There are many others. I am not a detail person and desperately needed help with this project. She spent untold hours getting permission to use some of the illustrations in this book as well as proofing and revising the final version of the manuscript more than once. She also followed the detailed format required by the publisher. A late comer to this project is Kayla Warner, a local university student and English major. She has helped with the references and formatting the entire manuscript. Thanks to all of you for your tireless help in making this lifetime dream become a reality.

Finally, many friends have encouraged me to complete this project. At times I was overwhelmed by the task and must confess I have been working on it for over twenty years on this book. I may be many things, but I am neither fast nor organized. Thanks to all of you for prodding me to complete this project. Special thanks to Sean and Julie Williams and to Julie's parents, Gene and Oma Hicks for providing endless support, prayer and encouragement for me to finish this project. Thanks too for making me feel a part of your family. Sean and Julie have two beautiful children, Sean Todd and Savannah Rose and I have dedicated children's books to both of them. They are the delight of my life.

As with any major writing project, there will undoubtedly be errors and omissions. I am responsible for both and hope readers will be forgiving and contact me regarding such matters. My email address is DocGater@aol.com or you may write me at: Route 5, Box 216, Weatherford, OK 73096 USA.

Meet the Author

Norbert Smith And A Soul Mate.

E. Norbert Smith was born in Lebanon, Oregon in 1941. My parents divorced when I was six months old and my earliest memories are living with my mother and grandparents on the family farm in western Oklahoma where I still live. I married as a teenager, but we divorced after 31 long years in an unhappy marriage. I have two grown children, Weena and Jayson. Weena and her husband, Mack, live in Florida where she teaches remedial English in the public school. Mack retired from an Air Force career and recently finished commercial airline pilot training. I also love flying and have owned two airplanes. Their son, Brent, is serving our country as a Marine sniper. Jayson lives in Norman, Oklahoma and is preparing for a career as an emergency medical technician. This should suit him well as he has always enjoyed driving fast.

I was raised in a Christian home and attended Emmanuel Baptist church of which my mother, grandparents and other family members were charter members over seventy-five years ago. It was a small country church back then, but moved into town in 1960. I accepted Christ as my personal Savior in that church at the tender age of six and was taught to see God in nature and to respect all living things. My favorite childhood Christian song was: ***Oh, the B-I-B-L-E, yes that's the book for me. I stand alone on the word of God, the B-I-B-L-E***. That little song was prophetic as I have spent much of my life standing alone on the Word of God.

15

I have always loved wild animals and as a boy kept a variety of non-traditional "pets" including snakes, lizards, turtles, salamanders, crows, skunks, opossums, squirrels, chipmunks, gophers, tarantulas, black widow spiders, scorpions and a variety of other fascinating creatures. Many years later, as a zoologist, I used radio telemetry to study the cardiovascular response to fear of some of these same animals, no doubt working with offspring of those I enjoyed as a boy. More recently I had a pet boa constrictor for 23 years. It is now on display at the Oklahoma City Zoo making baby boas.

When my high school biology teacher came to the part in the textbook about evolution, she closed the book and read the first two chapters of *Genesis.* She was a fine Christian woman and her intentions were good, but her approach had the opposite effect on me. I knew what the Bible said and as a typical rebellious teenager wanted to know what she was hiding. That same day, I bought Charles Darwin's *Origin of Species* and became infatuated with evolution. Like many today, I was deceived by evolution's appearance of scientific credibility. I still believed the Bible and my faith was never in doubt, but for the next seven years, I thought evolution was the method God used in creation. As often happens to individuals and to some Christian denominations, I began doubting the literal six day creation and other important teachings in the Bible. While still in high school, I became interested in electronics. During my junior year I obtained an amateur (ham) radio license from the Federal Communication Commission. My novice call sign, KN5PHD was prophetic of things to come for I had a Ph.D. from the FCC. I built my first transmitter from old radio parts and talked to other ham radio operators in 23 states. In 1959, five days after high school graduation, I joined the Air Force with the promise of electronics school in part because of my knowledge of radio. I finished at the top of my class and worked in electronics four years in the Air Force. After my honorable discharged in 1963, I worked three years at Texas Instruments near Dallas, Texas. Following endless discussions of origins, a dear friend and home church pastor Paul Ehmer gave me a copy of *The Genesis Flood* by Henry Morris and John Whitcomb. I quickly realized the Bible can be trusted and is without error where it touches on science or history. It also became obvious that the scientific evidence supporting evolution had been greatly exaggerated by the press and especially in science textbooks.

During my sophomore year in high school I felt God was calling me to special service in the area of science and the Bible. I surrendered to God's will, but did nothing about it until after reading *The Genesis Flood* book. Soon I felt the need a formal education in order to have a voice in the growing origins debate. I gave up a promising career in electronics to pursue a bachelor's degree in biology at Southwestern Oklahoma State University in my hometown of Weatherford, Oklahoma.

While attending the university, I designed a radio telemetry system for tracking rattlesnakes and realized my background in electronics would open doors for me in animal research. Indeed it has done so over the years. I was open and honest regarding my doubts about evolution with my advisor, Dr. Hobart Landreth. It was he that obtained a major National Science Foundation grant to study rattlesnakes, but it was my radio transmitter that enabled him to complete the research. We camped out together and caught and studied to movements of hundreds of rattlesnakes. We worked closely the entire three years it took me to finish college. A month before my graduation in 1970 Dr. Landreth came to my house and in front of my family informed me that because of my rejection of evolution he would not recommend him to any graduate school. He assured me none of the other biology professors would recommend me either. I was devastated and for the first time realized how serious it was to doubt Darwin. Nevertheless, I was accepted for graduate study at Baylor University, but remained silent regarding my views of origins. I designed and used a sophisticated multichannel radio telemetry system to study behavioral and physiological thermoregulation of free ranging alligators in south Texas. I earned a master's degree in biology from Baylor University in two years and had several research papers accepted for publication as a graduate student. Alligators had recently been classified as an endangered species and research money was available. My research was supported by the National Geographic Society and other national funding agencies. I had an article published in the *Creation Research Society Quarterly* two months after my graduation with a Master's Degree in Biology in 1972. Upon this discovery of that article my major professor phoned me and told me in no uncertain terms that if they had known of my rejection of evolution I would never have been allowed to study at Baylor University. The message is clear. If you want to graduate, even from a Christian university, "Don't dis Darwin."

I continued studying alligator thermoregulation at Texas Tech University and earned a doctorate in Zoology three years later in 1975. I have studied alligators in south Texas for a total of fifteen years and captured, taken blood samples and released over two hundred wild alligators. The largest one I studied weighted over 750 pounds. Alligators are fascinating creatures and have been very good to me. Shortly after earning my doctorate I had another article published in the *Creation Research Society Quarterly* and once again my graduate committee found out about it. They were so outraged that they actually formed a committee to annul my Ph.D. They were not successful, but this underscores the risks for those in academia of doubting Saint Darwin. Religious persecution is indeed alive and well in America today, yet few are aware of it. We are truly at war and the dangers are real.

After graduation with my doctorate I taught a summer school course at Fort Hays Kansas State University. I then accepted a position at Rochester Institute of Technology in upstate New York and taught biology courses for a year. Their grants department helped me obtain funding from the King Ranch in south Texas to continue studying alligators. I had outstanding student reviews and was offered the department chairman position if I stayed.

I missed my home turf and friends so returned to my home state and taught electronics and life science courses at Northeastern Oklahoma State University in Tahlequah, Oklahoma. While there I published about 50 technical papers and had dozens of excellent research students. Again I received outstanding student evaluations and I was awarded the Outstanding Teacher award two years. I continued studying the cardiovascular response of wild animals to fear and published several papers with my research students. In 1978, the BBC filmed portions of my alligator research for the television documentary entitled: *A smile for the Crocodile.* It has been widely seen in the United States and Europe.

Due in part to that TV documentary I was invited as a keynote speaker to a major international biotelemetry conference at Oxford University and took three of my research students with me to England. Two of the students gave technical papers at that prestigious international meeting. One was offered a post doctoral position in Europe, but had to decline as she was only a sophomore. I also met several scientists at the event and some later came to the United States and completed research studies with me. In spite of these accomplishments I was denied tenure ending his teaching career. Needless to say I was devastated and returned to the family farm and I worked a variety of jobs including farm work and as oil field roughneck. I finally found a teaching position in a Federal prison and eventually worked at a community college teaching mostly pre-nursing courses. Needing more income for retirement, I became an over-the-road truck driver. Driving trucks was the adventure of my life and I still sometimes miss it.

I have enjoyed writing most of my adult life and have published dozens of popular electronic hobbyist and children's magazine articles including articles and stories in *Popular Electronics*, *Radio TV Experimenter*, *Highlights* and *Ranger Rick*. I have also published over one hundred technical scientific papers in diverse areas including: the behavioral and physiological thermoregulation of alligators and other reptiles, the cardiovascular response of wild animals to fear and the design and application of sophisticated multichannel biotelemetry systems. I felt a strong kinship with the late "crocodile hunter" Steve Erwin and at the request of the editor of *The Citizen Scientist* wrote a brief eulogy *Remembering Steve Erwin.* Here is a link to the article online: www.sas.org/tcs/weeklyIssues_2006/2006-09-22/feature2p/index.html. I have remained active in Creation science and publishing dozens of articles in the *Creation Research Society Quarterly* and other Creation journals. My most recent paper was published in an Australian journal in 2010. (See: **Which prey do predators eat?** *Journal of Creation* 24(2):75-77)

For the past fifteen years, I have led an online group dealing with science, evolution and the Bible. It was started as an outreach for New Life Church in Colorado Springs when I was a truck driver. It was their first online small group, but at last count there are now more than 100. I see it as a way to do overseas mission work without the expense or need for a passport. The group varies in size from month to month, but has approximately one hundred members worldwide. Central to the ministry is the website: www.GodofCreation.com. It provides helpful information to science students, professors and others dealing with science and Creation issues. I also recently taught a distance learning online course about Creation for Liberty University in Lynchburg, Virginia.

This is my fifteenth book and perhaps my most important. It is my life's work and ministry...my *opus magnum* and took over 20 years to complete. I have also written a series of colorful children's books about an alligator and his friends in south Texas. I have started twelve more books that should be completed and available soon. All my books are available over the Internet, ask your favorite bookstore to get them for you or contact me. For the children's books search for "Al-the-Gator." The covers of the books can be seen online and are indicative of the quality of the drawings inside each book.

My first book*, Passive Fear: Alternative to Fight or Flight* was published in 2006 and is about my twenty years of using radio telemetry to investigate the cardiovascular response of wild animals to fear. I discovered the passive fear response by accident while working with alligators. I was the first scientist to understand the classic fight or flight response is only one option an animal has when approached by a potential enemy. Alternately, animals may hide or even feign death when frightened, as in the case of our American opossum. This remarkable and largely unknown passive response to fear is as widespread and every bit as important for survival as the better known fight or flight response. It is characterized by a marked decrease in heart and respiration rates reflecting a reduction in metabolism. The response is even mentioned in the Bible. The Roman soldiers were the best trained and most aggressive fighters at the time. Yet, upon witnessing the resurrection of Jesus Christ we are told, *The guards were so afraid of him that they shook and became like dead men.* (Matt 28:4, NIV) The opossum study was done with Dr. Geir Gabrielsen whom I met while lecturing at Oxford University. He just finished a post doctoral study at Harvard Medical School with infants at risk for Sudden Infant Death Syndrome (SIDS) and came to Oklahoma to study death feigning in opossums with me. We demonstrated even though they show do not show the normal eye blinking response when the cornea is touched, they remain fully conscious and reduce their heart rate again with the return of a dog or man. After our study he returned to

Harvard and showed his professor the following photograph of an opossum "playing dead."

Death Feigning in the Opossum.

Upon seeing this photograph his professor became excited and said, "That is just like babies with SIDS." They immediately went it to the hospital wing that contained babies at risk for SIDS and the doctor clapped his hands near the sleeping infants. Approximately 90% of those babies had a SIDS episode that could have resulted in their death had not a physician been there to resuscitate them. This means at least sometimes the death of a baby with SIDS may be triggered by a startle reflex. Obviously something else is wrong as they should start breathing again. Our opossum discoveries helped increase our understanding of Sudden Infant Death Syndrome (SIDS). The application has saved the lives of hundreds of babies and has been applied in other important areas. The book and several excellent reviews are available at Amazon.com. You can read the first chapter at: www.NorbertSmith.com.

My second book, Al-the-Gator and Freddy Frog is the first in a series of children's book and was published in 2007. It is a fictional book about an alligator in a farm pond in south Texas. Each book has some interesting biological facts and a list of additional resources. My first hand knowledge and research of alligators is uncommon for a children's author. The book and several reviews are available at Amazon.com by title or by my pen name "Doc Gator."

Battleground University is my third book and first Christian novel. It was published in 2008. It was an attempt to write a modern version of the classic C. S. Lewis Screwtape Letters and consists of letters from a demon attempting to influence two university professors. I completed the science portions of the book several years ago, but wanted to include a discussion of the dark side of the feminist movement. As a man, I felt it unwise to go there. Fortunately I met an online friend willing to contribute a contemporary view of not only the feminist movement, but many other relevant topics as well. She brings a fresh approach as a mother of four teenage daughters including a surprise set of beautiful triplets. Understanding the increasing dangers for Christians in the public schools, she home schooled her girls. She teaches at a major secular university, but feared getting fired if her school found out she had written a Christian book so she opted to use a pen name. This is sad, but wise in today's increasingly anti-Christian climate. As the book begins to sell a sequel is planned and perhaps a series. The setting in a major university provides unlimited character and plot possibilities. Again, for additional information and reviews check Amazon.com and my book website.

Al-the-Gator and Tommy Turtle and a companion coloring book are my fourth and fifth books and were published in 2009 by. They are the second and third in my series of children's books about an alligator and his friends. It seems children are fascinated by crocodiles and dinosaurs. The new character in this book is the red-eared slider turtle. Both books have many beautiful color illustrations as well as some scientific facts about the animals. A few references to technical papers I and others have written about the animals mentioned in the book are included for children that want to dig deeper. This book and those following have a few Bible verses about Creation and a moral lesson based on the story.

The sixth book (fourth children's book), Al-the-Gator and Annie Anhinga combines the story and activity books into one book and children seem to enjoy it. The activities are designed to help the children remember so of the important facts and moral lesson from the story. It was also published in 2009 and is available if you ask for it at your local bookstore, from the Internet or from me.

My fifth children's book was published in 2010 and is Al-the-Gator and Honey Bunny. The new character is a swamp rabbit. I studied swamp rabbits both in the laboratory and free ranging on an island in Lake Tenkiller in eastern Oklahoma for five years. Swamp rabbits are often found where alligators live and are excellent swimmers and divers. When forced underwater in a wire cage they showed the classic diving bradycardia response and reduced their heart rates dramatically. I also trained them to voluntarily dive under water for a carrot and they did so without any reduction in heart rate. As a result of the publication of this discovery, hundreds of other studies of diving animals had to be repeated under natural conditions and in each case what had been wrongly called "diving bradycardia" was instead a response to fear. Both alligators and swamp rabbits have been very good to me.

The reading material list also includes other books and articles I have written and reference material written by others. I feel strongly that children should be challenged to find and read material they find interesting. Even as a boy, I enjoyed reading "college level" books about animals. It was the world of reading that helped me develop a lifelong love for nature and all living things. It is why I became a zoologist. What child does not like learning about alligators and rabbits? Both are truly beautiful and fascinating animals…especially alligators.

My eighth book is American Buffalo and Native Americans. It has been accepted for publication by The Perfect Circle Publishing Company, a small Native American publisher in Oklahoma. It should be available early in 2011. The book is co-authored by my friend and Native American Dr. Jay Swallow. Central to the book is the scientific fact that bison are the only mammal with an incompletely divided mediastinum. This means that both lungs are functionally in the same cavity, making them easy to kill by a single arrow in the chest. Few people are aware of this important trait and it is impossible to defend from an evolutionary viewpoint as it has no survival value. It is obvious God created the bison special in order to provide food, shelter and clothing for our Native people. The book also contains some amazing Native American buffalo legends that have captivated our native plains people for centuries, yet few white people have any knowledge of these stories.

Another major book has been accepted for publication and should be out soon. It is Sacred Cows in Science, no Objectivity Allowed. This is the first book for which I only the editor and let me just say it has been an adventure as some people tend to procrastinate. Seventeen professional friends have contributed twenty chapters on topics about which they are passionate. This book is aimed at science teachers and science students. We are taught we have freedom of speech in America and that science is objective. Neither of those things is true. There are many sacred cows in science. The largest and most obvious sacred cow is evolution dogma. Anyone doubting evolution is openly ridiculed, denied tenure or fired. This has already happened to over 3,000 competent professors yet the media fails to report it. Freedom of speech does not exist when it comes to evolution. The public MUST be made aware of this increasingly prevalent form of religious persecution in America today. This book is finished and has been accepted for publication by WestBow Press, a Division of Thomas Nelson publisher and should be available early in 2011.

I have completed several more children's books and they should all be available in 2011. All of them are beautifully illustrated with whimsical color drawings and all have a moral lesson and list of additional reading material for those children that want to learn more. My tenth book and sixth children's book is Al-the-Gator and Sneaky Snake. The new character is the beautiful western ribbon Snake, a kind of garter snake found in South Texas where I studied alligators for many years.

The eleventh children's book is Al-the-Gator and Larry Lizard. The new character is the collared lizard or mountain boomer which is also Oklahoma's State lizard. Here is a photo of one living in my garden.

Beautiful Collard Lizard or "Mountain Boomer."

I enjoyed watching them as a young boy and as an adult attached heart rate radio transmitters and studied them with Ben Simard from Canada. As expected, when approached by man or a predator such as a dog, they run with increased heart rate. As soon as they reach the safety of their burrow and hide their heart rate drops far below pre-stimulus values. I enjoyed playing with these beautiful creatures as a boy and found it most rewarding to study them as a zoologist.

My twelfth book is **Gator Tales** about my dog and of the adventures we experienced over the years. Gator also helped with the opossum research by eliciting the death feigning response in them. It was written mostly for children, but adults seem to like the parts about truck driving. Gator grew up in an 18-wheeler. We all see trucks every day, but few know much about the trucking culture or what is actually involved.

My thirteenth book is **Al-the-Gator and Pattie 'Possum** and is about another animal I played with as a boy and studied extensively as a zoologist. I had the privilege working with Dr. Geir Gabrielsen from Norway on this study.

He and I met when I lectured at Oxford University and stayed in touch. He completed his doctorate in Norway and did a post doctoral study at Harvard before coming to Oklahoma to study opossums.

Al-the-Gator and Woody Woodchuck is my fourteenth book and ninth children's book to complete. This book is about yet another animal I studied for several years. I used heart rate transmitters to study the cardiovascular response to fear under free ranging conditions on an island in Lake Tenkiller, near Tahlequah, OK. I discovered several things about these fascinating animals, but unfortunately never did find out, *"How much a woodchuck could chuck if a woodchuck could chuck wood."* When they are feeding above ground and frightened, like most other animals they run with an increased heart rate. Upon retreating to the safety of their burrow their heart rate drops to about half the pre-stimulus values as I have discovered for many other animals. Of perhaps more interest is when a dog starts digging at the entrance of the burrow, the heart rate drops again to about half what it was after entering the borrow. Perhaps this is an expression of reduced metabolism and helps enable the animals to remain underground for an extended time and reduce the amount of oxygen consumed. As is often the case in scientific research more questions are raised than are answered. More work is needed.

The last of the children's books is *Al-the-Gator and Sophie Squirrel*. This is my tenth children's book. It was started and completed in a single week…a new personal best. It is about yet another animal I used radio telemetry to study its behavioral and cardiovascular response to fear under both captive and free ranging conditions. I have started several more books and they are in varying degrees of disarray.

After my tenure denial for accepting the Genesis account of Creation I eventually obtained a community college teaching position where I taught electronics five years inside a federal prison and later was able to teach life science courses on campus. I needed more income for retirement and became a truck driver. The first ten years I was an over-the-road driver and traveled over 2 million miles without an accident or traffic ticket. I also taught over 100 young drivers how to safely operate those big rigs. Truck driving was the adventure of my life and paid over twice what I could earn with a doctorate. I never tired of the endless change in scenery. I never missed teaching or the classroom, but still miss driving a big rig sometimes.

After ten years I finally got tired of living on the road and being away from my home and vegetable garden. I gave up my beloved job and turned in my truck. I still wanted to drive a truck, but wanted to drive for a local company so I could be home in the evenings and enjoy gardening once again. Near the terminal for the company I had been driving for was a sign for a local trucking company. I found the terminal and went in to ask a few questions. They were impressed with my safe driving experience and no tickets and offered me a job on the spot without even checking my references. It was a really good company to work for and I still return from time to time to visit with some of the drivers and staff.

If it doesn't have 22 wheels, it's not a real truck.

My new rig was a double tanker and I delivered fuel in Oklahoma City and western Oklahoma. I drove that rig until my retirement in 2005. It was an excellent company and my dispatcher was one of the best bosses I have ever worked for. I love retirement and now spend most of my time writing, gardening and online. I am still active in church and often speak about Creation and evolution at churches and university classrooms. Life is good and retirement made it even better.

Introduction and Overview

We live in a world where political correctness and tolerance to different worldviews are loudly proclaimed, yet contradictions abound. Children must recite Muslim prayers in many public schools; but praying to the Living God is strictly forbidden. Many people feel political correctness played a major role in failure to avoid the terrible Fort Hood massacre of November 5, 2009 that killed 13 adults and one unborn child. The terrorist, American-born Major Nidal Malik Hasan, was a known Muslim who stated his strong opposition to our wars in Iraq and Afghanistan several times, yet nothing was done to deter him in large part out of fear of being politically incorrect.

There was another recent example of intolerance in 2010 that received lots of news coverage on Fox News. Juan Williams is a well known journalist and political commentator for Fox News. He was ostensively fired from National Public Radio (NPR) News for voicing something most of us feel...that it made him a bit uneasy when seeing Muslim passengers as he boards an airplane. He was even told by NPR's CEO that he should discuss this with his psychiatrist. Many feel that was only an excuse to fire him and the real reason was his role as a conservative political commentator on Fox News.

The ever-popular Harry Potter books are touted by public school leaders as a boon to teaching children to read. Elementary and high school students are given witchcraft assignments on the Internet where they learn details of the occult and Satan worship, yet a military chaplain faced court martial for praying in the name of Jesus. Inconsistencies are prevalent. Another example occurred in 2008 when two Black Panther members were outside a Philadelphia polling place dressed in military style uniforms and were accused of shouting racial slurs at prospective voters. One was brandishing a night stick, yet two years later no legal action has been taken.

Something is radically wrong with our culture today. It advocates tolerance, yet has become highly intolerant to the Christian faith or anyone criticizing evolution. Christian values are under attack as never before by the courts and popular media. Popular television programs often show Christians and especially pastors as uninformed crackpots or worse. It is not a good time to be a Christian.

The Ten Commandments, revered by our Founding Fathers and considered by historians as the source of much of our United States Constitution, are being banned from schools and removed from public display. There is growing pressure to remove "Under God" from our Pledge of Allegiance and "In God We Trust" from our coins. Secular humanism has replaced Christianity as our nation's religion and naturalistic evolution is the foundation on which it firmly rests.

Evolution is taught as established fact in public school and university classrooms. Discussion of contrary views by dissenting scientists or mention of the lack of supporting scientific evidence is strictly banned. Although not reported by the media, religious persecution is becoming commonplace in the United States. Thousands of public school teachers and university professors have been denied tenure or promotion or even fired for doubting Darwin. Sadly, any reference to Intelligent Design or the Creation account in the Bible is strictly taboo. Over two hundred graduate students have been denied access to higher education for accepting Biblical truth, while being told science is above all things objective. It is not.

Recent surveys find less than 20 percent of university students believe there are any absolute moral values. Morality, they believe, depends on the circumstances and the individual. Situation ethics rule. As they said in the 1960's they are saying again, "If it feels good, do it." Predictably, there have been increases in date rapes, sexually transmitted diseases, abortions and shootings on high school and university campuses. Learning, as indicated by scores on national achievement tests, is at an all time low. Many American high school graduates are functionally illiterate. At the college where I last taught, fully 85 percent of incoming high school graduates were required to take 1-2 years of remedial courses, before they could enroll in our watered down "college" classes. Rampant immorality and failing public education account for the marked increase in home schooling and enrollment in Christian schools.

Now, as never before, there is need for teaching absolute truth. The Bible is unequivocal. Some things are wrong under all circumstances. Sin and judgment are real. Morality is absolute and does not depend on what one believes or rejects. Actions have consequences. One goal of this book is to provide pastors, teachers, students and others with a compilation of scripture references to God as Creator and Sustainer of the universe. Another goal is to provide discussion about our contemporary culture wars relating to evolution, science, creation and the Bible. The Bible has a great deal to say about creation and science as it does about human behavior and morality. Morality is not determined by a majority vote.

Evolution has had a terrible corrupting influence on society for 150 years. It is time its sordid history is exposed…no matter the consequences. The release of the Ben Stein movie, *Expelled: No Intelligence Allowed* increased public awareness and discussion of the resurgence of religious persecution caused by evolutionism. Dr. Bergman's soon to be published six volume series starting with; *Slaughter of the Dissidents* will provide documented case studies of hundreds of individuals who have suffered religious persecution in modern

America. Some of the problems I had in college are outlined in volume one. The second volume, ***Silencing the Dissidents*** is finished and will soon be available from Amazon.com or your local bookstore. In it he details my own tenure denial along with several other people that have lost their careers for simply rejecting macroevolution.   I strongly recommend both of these excellent books.   As educated caring adults we must be informed and inform others. These books will provide depth and continue where ***Expelled*** stopped.   Each book will be a powerful resource and the complete series should be in every church, public school and university library in the land.   As a nation, we must return to the Biblical foundation and Godly principles on which this country was established.

There is no reason to apologize for the science found in the Bible or for the growing interest in the Creation and Intelligent Design in the United States and worldwide. Literally thousands of highly trained and qualified scientists have rejected macroevolution.   Thousands have signed Jerry Bergman's Darwin's Skeptic list. (Visit: www.rae.org/darwinskeptics.html)   More are adding their names each month. Informed people no longer accept evolution dogma for two important reasons: the continuing lack of scientific evidence supporting macroevolution and its failure to provide a plausible mechanism for the origin of life and complexity of all living things. The stakes are high. Our country is in need of spiritual revival as never before. Indeed, it appears to be tittering on the brink of unparalleled godlessness. We must not only know what we believe, but why we believe it.

Of even greater importance, we must be able to defend our worldview and share it with others. Scripture demands nothing less. ***Always be prepared to give an answer to everyone who asks you to give the reason for the hope that you have. But do this with gentleness and respect.*** (1 Peter 3:15) It is hoped this book will provide reasons for accepting Biblical truth and rejecting evolution. As Christians we have no choice, but to share our faith with the world. ***Whoever acknowledges me before men, I will also acknowledge him before my Father in heaven. But whoever disowns me before men, I will disown him before my Father in heaven.*** (Matt 10:32-33) All scripture references are NIV unless indicated otherwise.

Overview

Following is a chapter-by-chapter summary of this book providing the reader with an overview and relating it to the current origins debate. It is hoped this material will provide useful, refreshing and pertinent information about this important issue. My prayer is many will be forever changed as it is shared and that people will again have absolute confidence in the truth and accuracy of God's inerrant Word.

Chapter 1: History of Science and Creation

This chapter introduces this important topic and puts the current origins debate into historical perspective. Many of the founding fathers of science had a deep abiding faith in God and accepted the Biblical account of creation as simple historical fact. They unashamedly held a Christian worldview, yet this distinction is no longer taught to students due to the influence of political correctness and today's history revisionists. We must remember our roots. Much of the foundation of science was discovered by men and women of faith who accepted scripture as divinely inspired and factual. There has been a recent explosion of interest in Intelligent Design and Creation science. The more we learn about the complexity of all living things, the less tenable is the origin of life by random chance.

Chapter 2: Creation in the Bible

The Bible contains 597 verses dealing specifically with God as Creator. Anything that appears that many times in Scripture is obviously important and must not be ignored. Supernatural creation is foundational to the Christian faith. Creation was a miracle. It was not a gradual slow undirected process as many today believe and as our students are taught from their science textbooks and in our universities. Without the original perfect creation and subsequent fall into sin, there is no need for Jesus Christ as Savior. These passages provide information about God as Creator and gives Christians reasons to lift our hearts and hands in honor of our Creator as we enjoy the endless wonders of nature. Scientists and others failing to see this are without excuse for as the Psalmist said, *the heavens declare the glory of God* (Psalms 19:1). This chapter is divided into two parts listing 554 verses from the Old Testament and 43 verses found in the New Testament that deal specifically with Creation.

Chapter 3: God Remains Active in Creation

There are 1,002 Bible verses proclaiming God continues to be active in our world today. Many Christians accept some form of theistic evolution and believe God directed or created the universe in the beginning, but then stepped back and is no longer active in that Creation today. This popular view is not supported by science or by scripture. The Bible clearly teaches that God has remained active in Creation throughout history and remains so today. *And He is before all things, and in Him all things consist.* (Col 1:17, NKJ) God did far more than wind up the universe and let it go. He continues to be actively involved with His Creation and with humankind. *Are not two sparrows sold for a penny? Yet not one of them will fall to the ground apart from the will of your Father. And even the very hairs of your head are all numbered* (Matt 10:29-33). Does this sound like a God that is unconcerned about His image bearers?

Chapter 4: Scientific Evidence For Creation

The scientific evidence supporting creation and the global flood is ignored today by science textbooks and by high school teachers and university professors. Students, pastors and others must consider the overwhelming scientific evidence supporting these historical events. The very fact that we have a rich and varied fossil record is unmistakable evidence of the global Flood described in God's Word. No one needs to apologize for the Biblical account of Creation and the Flood, for both are supported by a wealth of scientific evidence if one merely looks with an open mind. It is hoped this book will shed light on these important scientific evidences. We are taught science is above all, objective, yet it sadly falls far short of this oft stated goal.

Chapter 5: Science in the Bible

This chapter highlights some of the science mentioned in the Bible. Unlike every other book, the Bible claims divine authorship. Certainly, the Bible is important regarding spiritual matters but, if it is the infallible Word of the Living God, where it touches on history or science, it must also be correct. Because God is the Author of both science and His Word there can be no contradiction. There are many profound insights into modern science in the Bible written thousands of years ago by people with no knowledge of science. Many of the Biblical statements were diametrically opposed to conventional wisdom of the time. This is powerful evidence of the divine authorship of the Bible. It is hoped the discussion of science in the Bible will increase confidence in all scripture for the reader as it has for the author.

Chapter 6: Secular Science and the Bible At War

There are many obvious conflicts between modern secular science and the Bible. Science is part of man's God-given dominion over the earth, but like other gifts it has become tainted by the world. Sometimes conflicts occur due to a misinterpretation of science or scripture or both. Some of these areas of apparent conflict are discussed with possible solutions. Professors in science classes, even at major Christian universities, often dismiss the Biblical writing about Creation and the Flood as myth and say it is filled with errors and must therefore not be taken literally. Science students must be well grounded in Biblical truth and understand what the Bible teaches and understand why it can be accepted as Truth. As students they are at risk defending a Biblical worldview, especially in science classes. The academic environment is often hostile to the Christian worldview because we live in a lost and dying world. After earning their terminal degree they can then share their faith with confidence to those around them. They will become the "salt and light" as Jesus commanded (Mathew 5:13, 14).

Chapter 7: Scientific Evidence For Evolution

Science has accomplished many important things but, science is not infallible. It has been wrong many times in the past and will be wrong in the future. Hubris and arrogance abound among many scientists and they often take any attack on evolution as personal. We *must* be free to examine the evidence no matter where it leads. The actual scientific evidence for evolution is surprisingly weak, in spite of what we read in science textbooks and are repeatedly told by science teachers and the media. A discussion of the direct and indirect evidences used to support evolution is given. Some of the evidence involves circular reasoning. False and even fraudulent arguments have been used in textbooks for decades. Some of these are exposed and discussed in this chapter.

Chapter 8: Evolution in the Twenty-first Century

In this chapter we examine the current state of affairs of evolution with admissions of some of its weaknesses by leading evolutionists. The Dover Trial and its implications are discussed as well as a recent tax payer funded public broadcasting documentary on the subject. There remains a lack of scientific evidence supporting evolution, yet objectivity is lacking when evolution is discussed. It seems many people have their minds made up, no matter what evidence is discovered. It can be argued that evolution is held more as a relgion than as science. Much of it is based on faith and not scientific evidence. Objectivity regarding evolution is pitifully absent. It is increasingly obvious scientists no longer have the freedom to follow the evidence if that evidence leads to Intelligent Design or the Creator. It is for these reasons that thousands of highly trained scientists have rejected evolution as a plausible explanation for the origin and diversity of life. Still evolution dogma exists, but its numbers of supporters is dwindling rapidly. Soon it will be but a footnote in the history of biology.

Chapter 9: Theistic Evolution

This is a difficult, but important chapter for it deals with a topic that is widely accepted in many Christian circles and by several major Christian denominations and even some respected Bible seminaries. It is hoped readers will approach this chapter with an open mind as it deals with the important and controversial issue of theistic evolution. It seems that many espouse this view without fully understanding what is involved. They do so without an in depth study of what the Bible clearly states about this important issue or what evolutionists say on the topic. Many Christian leaders and denominations have started down this path only to find it leads to a dead end. They find themselves doubting larger and larger portions of God's Word. I know this from personal experience for I too, briefly trod this road. Readers will see the inherent dangers for individuals and denominations in their rejection of the account of Biblical Creation. Creation was a supernatural miracle and cannot be explained by science as a natural process.

Chapter 10: Evolution As Idolatry
The title of this chapter may seem a stretch to some. The idea that materialistic evolution is a modern form of idolatry is foreign to many. Few pastors mention it. Most Christians today see idolatry as something that occurred thousands of years ago in other parts of the world. Yet we hear indirect reference to it repeatedly on TV nature programs. Nature and more specifically, natural selection are often said to be "creative" or "clever." Such traits often accompanied by reverence, are given to the creature or more specifically to natural law instead of the Creator of all creatures who also established natural law. It is hoped this discussion will provide insight into this important issue for many are deceived. The Apostle Paul gave this stern warning: ***They exchanged the truth of God for a lie, and worshiped and served created things rather than the Creator-- who is forever praised. Amen*** (Rom 1:25). Again the words of Paul are insightful. ***And for this cause God shall send them strong delusion, that they should believe a lie*** (II Th 2:11, KJV). That lie is evolution and many have been deluded.

Chapter 11: Intelligent Design
This viewpoint is growing in popularity and is based in large part on the inability of evolution to explain the origin of life and the many complex structures and functions found in all living things. It is not motivated by scripture and is less offensive to many in science compared to the older Biblical Creation arguments. Most importantly the ID movement has spawned numerous *IDEA* clubs on high school and university campus all over our country where evolution, Creation and Intelligent Design are freely discussed. Such discussions are needed and blind eyes are being opened.

Chapter 12: Influence of Evolution On Society
This is perhaps the most important topic of discussion regarding origins. Evolution has exerted hideous societal influences past and present. The truth is not pretty. Evolution was justification for the horrific Jewish death camps of Germany and for "death with dignity" arguments for euthanasia today. It plays a

major role in the widespread practice of abortion. This may well be the most controversial chapter in the book, but I would be remiss to ignore the historical and current impact of evolution and have attempted to document the material presented. Jesus warned us. *By their fruit you will recognize them. Do people pick grapes from thornbushes, or figs from thistles? Likewise every good tree bears good fruit, but a bad tree bears bad fruit. A good tree cannot bear bad fruit, and a bad tree cannot bear good fruit. Every tree that does not bear good fruit is cut down and thrown into the fire. Thus, by their fruit you will recognize them.* (Matt 7:16-20) The fruit of godless evolution has been death to tens of millions of adults and unborn infants.

Chapter 13: Christian Persecution in America

This chapter illustrates how dangerous it is today to question godless evolution. Included are examples of highly qualified professors who have been denied tenure or fired for rejecting evolution. This I know from personal experience as do hundreds of others. Some of my problems in college are discussed are some of the details of my own tenure denial. The risks of doubting Darwin are great and continue to grow. Science is no longer objective and our very freedom of speech is under attack at our universities and elsewhere.

Chapter 14: Implications For Science

If Creation is true and evolution is false, there are many profound implications for science, for individuals and for society. Some of the implications of a Creation worldview and the scientific implications are discussed. Science has become a god to many in our world today, but all this changes if, in fact, the Living God created our world as the Bible clearly teaches. In spite of what we have been taught, scientists are often not objective. All of us see the world through the tinted lens of our worldview, including scientists. There is need for a major re-evaluation of science and science education. The stakes could not be higher. Not only are moral issues involved, but for science to remain productive, objectivity must return and be practiced, no matter where the evidence leads. Science, as we know it, will die if objectivity remains suppressed as it is today.

Chapter 15: Why Does It Matter?

This final chapter is an attempt to explain how resolving the origins issue is more important than mere scientific speculation. It is of eternal importance and it has implications and profound influence on people and society. I realize this topic is not a central issue for most people, but for many in science it is truly a heaven or hell issue. If living things, including man, were not the result of random variation over time, this changes everything. What if, as the Bible clearly proclaims an all powerful God breathed the universe and all living thing into existence. What if we really were made in the image of this all knowing and loving God? If this is true, we are responsible for the choices we make. What if there really is life after death and a real heaven and hell as the Bible clearly teaches? Souls hang in the balance for all eternity. Morality is absolute and not open for discussion. Judgment is unavoidable *for it is appointed for men to die once, but after this the judgment* (Heb 9:27, NKJ). Indeed, if the Bible is true and the scientific evidence actually supports Creation then everything changes. Carefully ponder the words of those who have gone down this path before you. Let us begin the journey down the road less traveled. If Creation is true and macroevolution is false, the ramifications are far-reaching and have eternal significance. This changes everything.

Chapter 1
Brief History of Creation Science

The heavens declare the glory of God; the skies proclaim the work of his hands. Day after day, they pour forth speech; night after night, they display knowledge. There is no speech or language where their voice is not heard. (Ps 19:1-3)

University professors, television nature programs and the media repeatedly tell us evolution is the single most important unifying principle in biology. Anyone with the courage to raise questions about the lack of supporting evidence is openly ridiculed and considered ignorant. They are accused of being religious zealots and are disparagingly compared to those who think the earth flat. The self-proclaimed "educated intelligentsia" seems blissfully unaware of the actual history of science. Many of the giants in science, including Newton, Kepler, Boyle, Maxwell, Faraday, Pasteur and a host of others, accepted as scientific fact, "In the Beginning, God created." Some felt privileged to be "thinking God's thoughts after Him." Sadly, science in the twenty-first century must be increasingly interpreted within the straightjacket of Darwinism. Mention of the Creator or discussion of the growing scientific evidence for design is strictly forbidden. Scientists no longer have the intellectual freedom to follow the evidence wherever it leads. The old adage "science vs. religion" is untrue, yet widely touted as gospel. True science and Bible-based views of origins are fully compatible. It is naturalistic macroevolution that is at loggerheads with both scripture and the actual scientific evidence.

Secular science is not the all-powerful worldview it was assumed to be throughout most of the nineteenth and especially the twentieth centuries. For example, the field of evolution-based science has not yet been able to identify the origin of the human mind, rational thought, free-will, or how an undirected explosion billions of years ago could produce thinking humans with the ability to contemplate their own origin (O'Leary, 2004; Beauregard and O'Leary, 2007). Nor has it proposed a rational explanation for the origin of life and the complexity found in all living things. Mutations are errors and most are harmful. They can not be creative. That is why we try to avoid mutagenic agents. These are but a few of the many questions evolution has failed to answer. Is the synthetic theory of evolution or neo-Darwinism the only way to address the broad and sometimes cryptic field of science? Hardly. Science is at times fallible with scientific explanations constantly being reevaluated and redefined. This was true in the past and remains true today. Consider, for example, recent observations that undermine long held notions about black holes (Robertson and Leiter, 2002, 2004, 2006; Schild, Leiter and Robertson 2006, 2008). More recently, fraud and outright dishonesty have been found to be rampant regarding the Global Warming charade.

Natural man with his vices, biases and preconceived notions ensures science is not as true and objective as secular humanists would have us believe. We all see the world through the tinted glass of our worldview. For far too long, the physical and especially the life sciences have been tainted by the leaven of secularism. A heavy price has been paid by this corrupting influence affecting how people view God, themselves and their place in the universe. It has also had a devastating influence on how humans treat each other and the unborn. Biblical and secular history as well as modern society have an important role regarding the direction, influence and limitations of science. With the rise of the modern creation science movement, and more recently, Intelligent Design, people no longer have to don secular glasses in order to observe natural phenomena or conduct experiments. Today, many are attempting to truly follow the evidence, although there remains strong resistance to this view in the ivory towers of academia and in the ever more secular and anti-Christian media.

The history of science is clearly linked to the current ID/evolution controversy. As we shall see, Bible-believing scientists have made some of the most significant scientific discoveries in the past. True science has always been consistent with observation and with the Word of the Living God for He is Author of both. Informed truthful secular opponents cannot continue to ignore the rich heritage of non-Darwinian science. Scientists must be free to view nature from a Biblical context and return science to its God-honoring position as part of man's dominion over creation as it was throughout much of history. Scientists must again have the freedom to follow the evidence, even if that evidence leads to design and ultimately to The Designer. Today there are over five thousand well qualified and published scientists that have serious doubts about the truth or adequacy evolution as an adequate explanation for the origin of life and diversity of living species. Over three thousand have willingly placed their names on a public list of those doubting Darwin and the number continues to grow. See Jerry Begrman's website: www.rae.org/darwinskeptics.html. Thousands more have such doubts, but feared repercussions if it became known. Indeed, it seems the very foundation of evolution is starting to crumble. Former evolutionists have abandoned evolution dogma like rats from a sinking ship. Many of these scientists are not Christians and have no religious motivation or agenda. They are simply searching for truth. Still, there are often devastating repercussions for taking such a heretical view (Bergman, 2008). One of the strongest advocates of

Intelligent Design today is William Dembski. He was fired from Baylor University for rejecting evolution and now works for the Discovery Institute. His best known ID books are The Design Revolution and Intelligent Design. Both are excellent reading and provide background for this exciting and provocative new frontier.

Creation Scientists of the Past

Creationists believe true science, correctly understood, is compatible with Scripture. One need only read of the great Bible-believing scientists of the past to see the validity of this statement. As mentioned above, many of the giants in the history of science including Newton, Kepler, Boyle, Maxwell, Pasteur, Faraday and dozens of others accepted the Biblical account of creation found in Genesis as historical fact. It can be shown that a vast majority of scientists prior to Darwin accepted the Biblical teaching that in the beginning was God alone. We must be aware of our past and learn from it. Perhaps this is more true in science than in

Astronomer and Mathematician, Johannes Kepler.

other areas because much of this rich history is ignored today. Let us now consider some of these great founders of modern science and their relation to the God of Creation in some detail.

Astronomer Johannes Kepler (1571-1630) was a recognized mathematician when a well-known astronomer saw how exceptional the young man was. Kepler was invited to join a group of respected astronomers. Up until Kepler's research, the planetary paths seemed complex and confused astronomers for decades. Such journeys across space required someone who could make sense of the planet's apparent complicated paths. In 1600, Kepler joined the astronomy group at the observatory in Prague. He stated that in his research, he was simply *"thinking God's thoughts after Him"* – a phrase that has been used often by scientists who are also believers.

Soon, this gifted man formulated the laws of planetary motion as we know them and made major contributions to mathematics. Such advancements would later provide Sir Isaac Newton with information regarding universal gravitation. Kepler aided navigators by publishing data on the positions of planets and stars and made many other contributions to science. For example, Kepler discovered what would later prove to be a supernova, improved the telescope and the field of optics, and contributed to better understanding eye function. Kepler was also a

Mathematician and Philosopher, Blaise Pascal.

dedicated missionary-minded Christian and personally financed the translation of the Bible into Arabic, Irish, and Turkish.

Blaise Pascal (1623-1662) was a deeply spiritual man. He was the father of hydrostatics, a founder of hydrodynamics, and a great mathematician and philosopher. His mathematic contributions include probability theory and the modern treatment of conic sections. He would be branded today as politically incorrect for what is widely known as "Pascal's Wager" saying simply that his skeptical friends could lose everything in hell while Christians had nothing to lose, and heaven to gain for believing in God. Pascal is well known for his religious contributions, in particular his classic *Pensees*. – A defense of the Christian faith.

Robert Boyle, Founder of Modern Chemistry.

Robert Boyle, (1627-1691) was an Irish chemist, physicist and inventor. He is considered the founder of modern chemistry. He is best known for the formulation of Boyle's law relating the pressure and volume of ideal gases. His **classic *Sceptical Chymist*,** published in 1661, is seen as an important foundation for chemistry and introduced the concept that matter consisted of atoms and clusters of atoms in constant motion. He advocated all scientific theories must be proved by experiment before they could be considered true.

He was a dedicated Christian and contributed liberally to missionary organizations and to the translation of the Bible. He was motivated to study science because from ***knowledge of God's Work we shall know Him*** (Strobel, 2005). He founded the Boyle lecture series, intended to defend the Christian faith against those he considered "Notorious infidels, namely atheist, deists, pagans, Jews and Muslims" with the restriction that controversies between Christians were not to be mentioned. Since 2004, new Boyle Lectures have been given annually in London at the parish church of Saint Mary-le-Bow. The lectures, by an invited distinguished theologian or scientist, are designed to address topics that explore the relationship between Christianity and our current understanding of the natural world.

Sir Isaac Newton

Sir Isaac Newton (1642-1727) was arguably one of the greatest scientists who ever lived and bluntly stated that *all* of his discoveries were made in an answer to prayer. *We account the Scriptures of God to be the most sublime philosophy. I find more sure marks of authenticity in the Bible than in any profane* *history whatsoever.* – Sir Isasc Newton (Thayer, 1953). This important aspect of Newton is totally ignored by professors and textbooks today.

Newton also stated, this most beautiful system of the sun, planets, and comets, could only proceed from the counsel and dominion of an intelligent Being. This Being governs all things, not as the soul of the world, but as Lord over all; and on account of his dominion he is wont to be called "Lord God" or "Universal Ruler." The Supreme God is a Being eternal, infinite, absolutely perfect'. Opposition to godliness is atheism in profession and idolatry in practice. Atheism is so senseless and odious to mankind that it never had many professors (Thayer, 1953). How times have changed!

Carl Linnaeus, Father of Modern Taxonomy. Photo public domain.

Carl Linnaeus (1707-1778) also known as Carl von Linné or Carolus Linnaeus, was a Swedish scientist, Christian and creationist who put together the biological classification system of animals and plants that remains in use today. He is considered the father of modern taxonomy and held in high esteem by biologists all over the world. He believed the Genesis account of creation and saw the entire living world as created by the Living God. Consider his praise of God through Creation. *The flowers' leaves...serve as bridal beds which the Creator has so gloriously arranged, adorned with such noble bed curtains, and perfumed with so many soft scents that the bridegroom with his bride might there celebrate their nuptials with so much the greater solemnity* (Linnaeus, 1730). Linnaeus authored 180 publications regarding plants and their diseases.

English Chemist.

Michael Faraday (1791-1867) was born the same year as Samuel Morse, and like Morse, Faraday was a devout Christian and man of science. He is remembered as the inventor of the transformer, generator and electric motor. Faraday made many significant contributions in the fields of electromagnetism (he introduced the concept of magnetic lines of force) and electrochemistry that two units in physics (faraday and farad) are named in his honor. Indeed, a number of science historians refer to Faraday as the best experimentalist in the history of science. Faraday was the first holder in the *Fullerian Professor of Chemistry* at the Royal Institution of Great Britain and was appointed to that post for life. In his native England in March of 1991, Faraday was honored with a special first-day cover and a commemorative postage stamp. In fact, his signature and portrait were to replace William Shakespeare's on England's twenty pound notes, quite an accomplishment! Special exhibitions of Faraday were held at the National Portrait Gallery, The Science Museum, and the Royal Institution. Even a special memorial service was held in the revered Westminster Abbey for this man of God and man of science.

42

Samuel Morse, inventor of the telegraph.

Telegraph inventor **Samuel F. B. Morse** (1791-1872) graduated from Yale in 1810. Unknown to many, he was a gifted artist and he made the world's first photographic portrait. Samuel was the son of Jedediah Morse, a Flood geologist and foremost geographer. He honored the Creator by sending the first telegraph message in 1844 as a quote from the Bible: *`What hath God wrought'* (Numbers 23:23).

French biologist **Henri Fabre** (1823-1915) was a popular teacher, physicist, botanist and friend of fellow Creationist, Louis Pasteur. He is known for his meticulous work with insects. He is considered the father of entomology, the study of insects. At nineteen, he earned a primary teaching certificate and later wrote ten books about insects, all without a giving so much as a nod to Darwin's theory. For this he was criticized by the secular establishment. He also wrote children's books conveying God as Creator and in later years received numerous awards for his research.

Entomologist, Physicist and Botanist, Henry Fabre.

James Joule, Father of Thermodynamics

James Prescott Joule (1818-1889) is known as the father of thermodynamics, demonstrating the first law of thermodynamics. Joule formulated what was to be known as Joule's Law of electricity which states the power dissipated as heat is equal to resistance times current squared. The unit of energy, the joule, is named after him. The most significant contribution was made in 1840 when Joule discovered the constant known as the "mechanical equivalent of heat." As is so often the case, Joule was a businessman first, making his discoveries in physics and thermodynamics as a hobby, but nonetheless received many honors in science. Joule was known as a man of sincere Christian faith. It's about time for the world to hear the wisdom of these words, because rarely has such a clear statement been given on why science should be the enthusiastic pursuit of the devout Christian. In his notes, Joule talks about many things: the value of science education for the youth, his opposition to science being applied to warfare or politics, the value of mathematical rigor, and the need for precision and planning in experimentation. He describes the ideal moral character of the scientist: one must be humble, diligent, energetic, prudent and zealous, pursuing science due to "a love of wisdom which unfolds, a love of truth for its own sake independently with regard to the advantages of whatever kind is expected to derive from it." Science and knowledge elevate us above the beasts that perish, and enrich our lives with "varied and fresh enjoyments." Among these random but uplifting thoughts, I will end with two quotes that so well express the theme of this book, that good science, the best science, is the fruit of devout love of God as Creator. To Joule, the study of nature and her laws is "essentially a holy undertaking," second only to worship as the rightful response to the Maker of all things. Hear the words of James Prescott Joule: *After the knowledge of, and obedience to, the will of God, the next aim must be to know something of His attributes of wisdom, power and goodness as evidenced by His handiwork. It is evident that an acquaintance with natural laws means no less than an acquaintance with the mind of God therein expressed* (Crowther, 1935).

Louis Pasteur (1822 -1895) was an amazing scientist and one of the greatest biologists of all time. He was awarded France's highest award for his remarkable efforts and discoveries (e.g. sterilization and pasteurization). His brilliant experiment with swan-necked flasks destroyed the myth of spontaneous generation, a foundation of evolutionism that he actively opposed. His work in medicine was well-known, developing vaccines for anthrax, rabies and diphtheria and formulating the germ theory of disease. Pasteur was victorious over an enemy he could not even see (the rabies virus)! He discovered three of perhaps

Inventor of Vaccines, Louis Pasteur.

the best known bacteria in microbiology:pneumococcus, streptococcus and staphylococcus. As remains true for many scientists in the twenty-first century, Pasteur saw no conflicts between Christianity and science. The son-in-law of Pasteur, Rene Vallery Radot stated that Pasteur had absolute faith in God and in eternity.

George Washington Carver, Plant Scientist Extraordinaire.

Famed plant scientist **George Washington Carver** (1864-1943) was born a slave in America, but went on to work his way through college in the north. He became a faculty member at the Tuskegee Institute in Alabama. Carver developed over 320 products from peanuts and 188 products from sweet potatoes. He was not only a chemist involved in agriculture, but also was an educator, showing an intense interest in his students at the Tuskegee Institute. For thirty years, he taught a Sunday evening Bible class and at four every morning would get up to pray. The inscription on the Roosevelt Medal he was awarded (1939) states, *To a scientist humbly seeking the guidance of God and a liberator to men of the white race as well as the black.*

Rocket Scientist, Wernher von Braun.

Wernher von Braun (1912-1977) was a German rocket pioneer, obtaining his Ph.D. from the University of Berlin. An active Lutheran, he wrote an essay on science and the Christian faith and, after becoming an American citizen in 1955, was the mastermind behind the Apollo program. *While the admission of a design for the universe ultimately raises the question of a Designer (a subject outside of science), the scientific method does not allow us to exclude* *data which lead to the conclusion that the universe, life and man are based on design. To be forced to believe only one conclusion—that everything in the universe happened by chance—would violate the very objectivity of science itself* (Neufeld, 2008).

There were many other notable scientists in the past and the reader interested in learning more about them should consult the myriad of excellent history of science books available at any good public or university library or online. The purpose of this introduction was simply to illustrate good science is consistent with a firm belief in the Bible and a Christian worldview. Much of the above information came from Dr. Henry Morris' revealing book, ***Men of Science Men of Faith.*** See this excellent little book for greater depth and many more examples (Morris, 1982).

Creation Science Today

When considering the history of Creation and evolution, one must keep in mind there has *never* been a conflict between true science and the Bible, for God is the Author of both. The idea of a conflict between science and the Bible has its roots in the naturalistic camp and is an attempt to portray creationists and to some extent all Christians as "anti-science." This is obviously not the case as we have seen with so many of the true leaders in science accepting the God of Creation and the historical truth of the Bible. Regardless of the facts, evolution activists continually present the false picture that science and religion are at war. They also assume such Creationist organizations as the Creation Research Society (CRS, www.creationresearch.org), Institute for Creation Research (ICR, www.icr.org), and Answers in Genesis (www.answersingenesis.org) and others are guilty by association.

Scientists in each of these organizations have advanced university degrees in various scientific disciplines from accredited secular institutions. In addition to their writing about Creation, they also do stellar scientific research and are published in major peer reviewed scientific journals. To set the record straight, creationists have never seen true science as an enemy, they embrace it! The accusation that we somehow fear science or find it in conflict with Scripture is patently false. We see science as part of the dominion God gave man over the creation in Genesis 1:26 and 28. Many of us see scientific research as our responsibility and our chosen career. Evolutionary humanists on the other hand, have for decades portrayed their position based firmly on atheistic natural secular humanism. *But man himself and his behavior are [sic] an emergent product of purely fortuitous mutations and evolution by natural selection acting upon them. Non-purposive natural selection has produced purposive human behavior*

(Haogland, 1964). Such non-scientific statements stretch credibility to the breaking point and beyond.

As Christians we must be neither surprised nor discouraged by such words for we cannot expect sinful man to see or accept the God of Creation. To the world it is but foolishness as the word of the Living God predicted long ago before the birth of modern secular science. *For although they knew God, they neither glorified him as God nor gave thanks to him, but their thinking became futile and their foolish hearts were darkened. Although they claimed to be wise, they became fools* (Rom 1:21-22) They see God in nature, but quickly turn away and prefer to believe the lie of godless evolution and live in darkness. This is human nature. King David said it clearly. *The fool says in his heart, "There is no God."* (Ps 14:1a) As it was then, so it is today.

Until the mid-nineteen hundreds, flood geology was accepted by most Christians and the science community, but had not been grounded in scientific research. This dramatically changed in the decades after 1961 and the publication of Morris & Whitcomb's 550-page tightly written *The Genesis Flood*. The influence of that profoundly important book is still felt today and is the primary reason for my writing this book, nearly a half a century later. That book forever changed the way I looked at science and the Bible as it has done for countless thousands around the world.

It was in September 1953 when Henry Morris visited Grace Theological Seminary where John Whitcomb was professor of Old Testament. Dr. Morris presented a paper to the American Scientific Affiliation entitled, *Biblical Evidence for a Recent Creation and Universal Deluge.* Because of this profoundly influential paper, Dr. Whitcomb drastically changed his approach to origins. During the next four years, he wrote his 450-page doctoral dissertation entitled, *The Genesis Flood: An Investigation of its Geographical Extent, Geological Effects, and Chronological setting* (Winona Lake Grace Theological Seminary, 1957).

This pivotal work would be the primary stimulus activating and uniting literally hundreds of creation scientists around the world. Henry Morris, in turn, was influenced by the flood geology work of George McCready Price (1870-1963) earlier in the twentieth century. Dr. Morris's life was forever changed as well. In his own words, *The publication of The Genesis Flood made a tremendous difference in my life, culminating in a change from engineering to full-time concentration on creationism and Christian evidences.* There were numerous speaking requests, then the formation of the Creation Research Society in 1963 and followed by the Institute for Creation Research in 1970. Soon there were numerous and extensive seminars, conferences, debate, etc., all over the world. Many have attributed the global revival of scientific Biblical creationism to the catalytic effect of that single book.

Christians involved in various scientific fields in the eighteenth and nineteenth centuries saw clearly that the unguided materialistic worldview could not explain the origin of the universe, the solar system or the origin, diversity and complexity of plant and animal species. A landmark meeting of the General Assembly of the Presbyterian Church (1910) produced the "five fundamentals," a condensation of the Christian faith. One critical fundamental was that Scripture was inerrant and this included the creation account in Genesis.

Through the decades, a number of Christian and Jewish scholars decided that Genesis was not to be taken literally and was instead poetical or allegorical. This was reinforced by the writings of Charles Lyell (1797-1875) whose book *Principles of Geology* influenced a young Charles Darwin. When the developing science of natural history began to be formed by men who denied Genesis (and with it the worldwide Flood), it goes without saying that there would be many scientific contradictions. Man began to depend on an atheistic or secular worldview, denying the Biblical record including the global flood. Not surprisingly, Darwin's idea of the origin of new species by natural selection was a further secular challenge to the Genesis account of creation. Today, many in the scientific community still cling to this failed idea and the origins debate continues today.

The teaching of Darwinian evolution in American public education became controversial after the First World War. Many people educated in the sciences were not convinced Darwinian evolution was true and saw the state as being active in foisting a secular anti-Christian philosophy on young people in public education. Prominent spokesman William Jennings Bryan (1860-1925) of the Democratic party saw what Darwinism was teaching using taxpayer dollars and voiced his opposition. The year Bryan died, G.K. Chesterton published *The Everlasting Man*, a book noted for exposing the logical flaws of Darwinism as well as coherently describing creationist ideas. There were other unsuccessful attempts to institute a formal witness for creation science. One was, *The Religion and Science Association*, founded in 1935. It only lasted only two years. Another failed attempt was *The Society for the Study of Creation, the Deluge, and Related Sciences* lasting from 1938 to 1945. These were unsuccessful largely because of the failure to unite those who wanted to accommodate the alleged geological ages in their worldviews with those holding to a Biblically based young earth and flood geology.

Shortly before the beginning of the twentieth century, Nietzsche loudly proclaimed God was dead or at least that the need for God no longer existed. There was a growing "Death of God" movement in the United States in the late 1950's and early 60's. The cover of *Time* magazine on April 8, 1966 proclaimed for all to see that God was dead. Even before this, in 1959, there was highly publicized convocation of secular scientists who met in Illinois to pay homage to

Charles Darwin and to hold a widely publicized graveside service for Christianity in general and the creation model in particular. The actually believed Christianity would soon be but a memory. It is obvious the funeral was premature.

The Darwinian Centennial at Chicago University marked one hundred years since the publication of Darwin's *On the Origin of Species* where many throughout the world assembled to bend their collective knees to Saint Darwin. He was eulogized for bringing the world "out from under the thumb of God" and into "evolutionary freedom." The keynote speaker of the Centennial was Julian Huxley, grandson of Thomas Henry Huxley the renounced atheist best known as Darwin's Bulldog, a staunch defender of Darwin's new theory . A committed humanist, who arrogantly declared secular humanism, would soon become the worldwide religion. *In the evolutionary system of thought, there is no longer need or room for the supernatural. The earth was not created; it evolved. So did all the animals and plants that inhabit it, including our human selves, mind and soul, as well as brain and body. So did religion. Evolutionary man can no longer take refuge from his loneliness by creeping for shelter into the arms of a divinized father figure whom he himself has created, nor escape from the responsibility of making decisions by sheltering under the umbrella of divine authority, nor absolve himself from the hard task of meeting his present problems and planning his future by relying on the will of an omniscient, but unfortunately inscrutable providence* (Huxley, 1959).

Just a few years earlier, graduate student Stanley Miller produced some sugars and amino acids in an enclosed apparatus. The media triumphantly and prematurely proclaimed that man had "created life" in the laboratory. The underlying message was clear, there was no longer need for God. Secular humanism had arrived and Man alone ruled the universe. However, in the twenty-first century the many problems with life-from-non-life have increased (e.g. an astrobiologist from Pennsylvania State University presenting evidence for high oxygen levels on earth "3.8 billion years ago" (Ohmoto et. al, 2006). Oxygen is the enemy of such scenarios because it is reactive and highly destructive to molecules necessary for life.) *By the start of the twentieth century, gradualism had achieved a state of complete ascendancy over catastrophism, a situation that was to persist for eighty years or more* (Palmer, 1999).

As these and other important events were unfolding, Henry Morris and John Whitcomb were writing *The Genesis Flood*. God used that profoundly important book to energize the revival of creation science in the last half of the twentieth century. The year 1980 became something as a defining moment as well. There was a slow resurgence of a previously ignored concept by the secular community regarding catastrophism. *Catastrophism is enjoying a renaissance in geology. For the last 180 years, geologists have applied consistently an uniformitarian approach to their studies that has stressed slow gradual changes as defined by Lamarck, Lyell and Darwin. Now, many of us are accepting that unusual catastrophic events have occurred repeatedly during the course of Earth's history. The events were significant, since they caused sudden drastic environmental disturbances as well as mass extinctions* (Hsu & McKenzie, 1986). The publication of *The Genesis Flood* and the rise of neo-catastrophism were profoundly harmonious. Prior to the release of the highly cited *Principles of Geology* by Scottish lawyer Charles Lyell, most of those involved in science accepted the Genesis account of Creation and the Flood as actual historical events. However, Lyell's book questioned such assumptions and proved to be his most important and influential, publication. This is mainly due to his introducing the philosophy of uniformitarianism: "the present is the key to the past." Lyell is seen by many as being the founder of ascribing "millions of years" to the sedimentary rock units that make up the "geologic column." For many decades onward, this idea permeated science crippling research, particularly in geology. Gradually there were seen significant problems that uniformitarianism could not address. *Furthermore, much of Lyell's uniformitarianism, specifically his ideas on identity of ancient and modern causes, gradualism and constancy of rate, has been explicitly refuted by the definitive modern sources, as well as by an overwhelming preponderance of evidence that, as substantive theories, his ideas on these matters were simply wrong* (Shea, 1982).

The resurgence of catastrophism, the process of catastrophic plate tectonics, and the resultant ice age just thousands of years ago provide a more rational alternative to the uniformitarian philosophy and are more compatible with scripture. The problems regarding Lyell's ideas were numerous, including the confusion he spread by deliberately providing two different meanings to uniformitarianism. This includes the massive extent of rock formations that required remarkable causes. There was also the problem in orology or the study of mountain formation for the latest evidence in showing their surprisingly young age. It is now generally accepted that a catastrophic collision of the Asian and Indian plates produced the Himalayas, including Mount Everest – which was not there before the flood.

Finally, many saw the mass extinctions of the past as not properly addressed by uniformitarianism's slow and small changes process spanning "millions and millions of years." Clearly, the flood described in Genesis was a catastrophe of global proportions. It caused a massive extinction never seen before or since. It clearly left its mark on the Earth. As predicated by flood geology, fossil beds thousands of feet deep are in the Earth's crust. Ensconced in these beds for all to see are trillions of terrestrial creatures, not just marine animals. Furthermore, these sedimentary fossil beds show clearly they were deposited by water currents on every continent. There can be no stronger proof of the global flood described in Genesis.

If the Earth were millions of years old, there would be indications of changed water current directions around the surface of the Earth due to different areas containing different sedimentary strata. Many catastrophists are not believers in the Biblical account of creation and the flood, but they do acknowledge a series of catastrophic events that shaped the world we see today. It would seem they are drawing away from uniformitarianism while getting closer and closer to a Biblical or catastrophist view the history of the Earth with of one massive catastrophe and catastrophic plate tectonics just thousands of years ago. Today, it would seem the drifting of the continents has stopped. *Meanwhile, since the continents drift as slowly as one's fingernails grow – from one to ten centimeters per year – even the most precise surveying methods available today have not yet detected drift* (Dietz, 1983).

Jesus Christ referred to the Noah and the flood as a real historical event. As it was in the days of Noah, so it will be at the coming of the Son of Man. *For in the days before the flood, people were eating and drinking, marrying and giving in marriage, up to the day Noah entered the ark; and they knew nothing about what would happen until the flood came and took them all away. That is how it will be at the coming of the Son of Man.* (Matt 24:37-39) Yet, because it is not supported by secular science, many professing Christians have chosen to reject the Genesis account as a literal historical record. Instead, they attempt to combine the atheistic philosophy of naturalistic evolution with scripture. One prominent group that does this call themselves "progressive creationists" led by Canadian astronomer Hugh Ross. They reduce the worldwide Genesis Flood to a local Mesopotamian flood in order to justify the alleged millions of years of sedimentation and fossilization *before* man's creation. Both science and scripture become insurmountable roadblocks to such a philosophy.

These two worldviews are incompatible; they will never find common ground because they are based on two opposing principles. Harmony is impossible. One worldview is supernatural, maintaining that scripture should govern our interpretation of scientific data; the alternative places secular view above scripture and proclaims the ever-changing majority opinion should dictate the interpretation of Genesis. The tenuous case for atheistic macroevolution has been repeatedly exposed by creationists as well as by those secular writers and scientists who remain skeptical of Darwinism.

The number of Darwin skeptics is growing explosively with thousands having the courage to sign the Darwin skeptics list mentioned above. Similarly, polls in the twenty-first century show that those holding to strict Darwinism comprise a mere 9 percent of the population. Interestingly, the percentage of those calling themselves non-theists is approximately nine percent as well. In contrast with the general population, the prestigious National Academy of Science looking at the philosophical convictions of scientists shows that approximately 95 percent of those polled do not believe in God (Larson and Witham, 1999). We can see that there is much needed change in the wind as science continues to uncover serious philosophical and scientific flaws in macroevolution.

On a related matter, the National Academy of Science has repeatedly flooded public education teachers with well written beautiful multicolor pamphlets assuring teachers there is no conflict between secular science and a belief in God. They continue to put down Creationists and anyone rejecting evolution as uneducated fools clinging to superstition and myth. Sadly, neither most teachers nor their students have the training and courage to stand up to such overwhelming pressure from such a respected group.

Philosophically, some Darwinists are beginning to take a hard look at their worldview. A popular example is atheist and philosopher of science Michael Ruse who was a key spokesman for the infamous 1980's McLean v. Arkansas trial, in which Federal Judge William Overton ruled that Arkansas' "Balanced Treatment Act" was unconstitutional. In a talk many years later, Ruse assured his perplexed audience that he was "no less of an evolutionist now than he ever was" but had given much thought to Phil Johnson's surgical exposure of macroevolution. Ruse came to realize that he, as an evolutionist, is metaphysically based at some level just as much are creationists...*I must confess, in the ten years since I . . . appeared in the Creationism Trial in Arkansas . . . I've been coming to this kind of position myself* (Buell, 1994).

In the 1980's, about twenty-five years after the publication of *The Genesis Flood* book, two important counter evolution movements appeared. Both of these new movements reject naturalistic evolution, but at the same time are also opposed to the Biblical account of Creation and creation science in general. On the downside, many have been deceived into abandoning the clear historical teachings of Genesis. One of these movements is Progressive Creationism and it will be discussed briefly below. The other is Intelligent Design and it will be covered in depth in a later chapter. The first of these counter creation movements is called "Progressive Creationism" advanced by Dr. Huge Ross, a Christian astronomer and others. He advocates creation started billions of years ago with the big bang and that animals were supernaturally created from time to time over millions of years. He rejects evolution and Adam's rebellion against God was not the cause of death. He sees the Genesis Flood as a local event and feels the sixty-six books of the Bible are fully inspired, yet must be understood in light of the sixty-seventh book, modern science. Perhaps the greatest weakness of progressive creation is its failure to explain the sedimentation and fossilization apart from a global flood. Again, we are reminded of the clear scriptural warnings of such erroneous teachings (2 Tim 4:3-4). That time has arrived.

The first of these counter creation movements is called "Progressive Creationism" advanced by Dr. Huge Ross, a Christian astronomer and others. He advocates creation started billions of years ago with the big bang and that animals were supernaturally created from time to time over millions of years. He rejects evolution and Adam's rebellion against God was not the cause of death. He sees the Genesis Flood as a local event and feels the sixty-six books of the Bible are fully inspired, yet must be understood in light of the sixty-seventh book, modern science. Perhaps the greatest weakness of progressive creation is its failure to explain the sedimentation and fossilization apart from a global flood. Again, we are reminded of the clear scriptural warnings of such erroneous teachings (2 Tim 4:3-4). That time has arrived. Next, we will look in depth at what the Bible has to say about creation.

Anhinga, Diving Bird.

Chapter 2
Creation in the Bible

In the beginning God created the heavens and the earth. (Gen 1:1)

Michelangelo's *Creation of Adam*

    The Bible is different from all other religious books. Other books are about man searching for God. The Bible is God's message to man and begins with His identification as Creator. God is many things, but few characteristics identify Him more clearly than that of Creator. He existed before the universe. He is first and foremost to be worshiped as Creator of all things. In the Old Testament, we see the character of God and his dealings with his ultimate creation and image bearer, humankind. We see His love and we see His wrath. We see His patience and wisdom. Above all, we see Him as our all powerful and always caring Creator. As Creator of the universe, He alone is worthy of our praise, adoration and worship. This is confirmed in the New Testament.

    Through him all things were made; without him nothing was made that has been made. (John 1:3)

    As a young man, I was amazed at the number of Bible passages dealing with Creation. For many years I thought the purpose of all those references was to give us information on how God created. I was wrong. The reason we have hundreds of such references is clear. As we view the marvelous details of the Creation we are to lift our hearts and our hands in worship of Him who made it all. Creation was not a process, but was a miracle…the miracles of miracles. Evolution is a delusion…a fairytale for those unwilling to face God. Peter warns us in these last days man will be "*willfully ignorant*" of the Creation. (2 Peter 3:3-7, KJV). We are also warned that men will be deluded. *For this reason God sends them a powerful delusion so that they will believe the lie.* (II Th 2:11) That lie is evolution. We must look at nature and see the hand of the Living God. *Through him all things were made; without him nothing was made that has been made.* (John 1:3)

Following is an attempt to list all Bible references relating to God as Creator. Passages from the Old Testament will be presented first followed by those in the New Testament. Any topic mentioned hundreds of times in scripture must be important, yet the Biblical account of creation is being denied by the world today as allegory or myth. This is true not only by secular science, but by many pastors and so-called "Bible scholars." Perhaps the greatest danger is the God of Creation is being denied today by an increasing number of Christian universities and seminaries. We must return to the God of the Bible and see Him once again as the all-powerful Creator. Without the perfect Creation and our fall into sin, there is no need for Jesus Christ as Savior. Creation is central to the gospel. As you consider the following passages, you will see why these truths are relevant and desperately needed in our world today. Let us return to our God fearing Christian roots as individuals, as churches and as a nation.

Old Testament Creation Verses

The following table provides a list of Creation verses by books in the Old Testament. Some will find it surprising that the most passages dealing with Creation are not found in Genesis, but in Job the oldest book of the Bible. It is also significant that nearly one hundred verses are found in the Psalms. The reason is obvious. We are to lift our hearts and hands in praise of God as Creator for He alone is worthy of such adoration.

454 Creation verses in the Old Testament

Book	verses	Book	verses	Book	verses
Genesis	58	Psalms	99	Nehemiah	1
Exodus	3	Job	179	Hosea	1
Deuteronomy	3	Proverbs	17	Joel	1
2 Kings	1	Ecclesiastes	5	Amos	13
1 Chronicles	2	Isaiah	43	Jonah	4
2 Chronicles	1	Jeremiah	22	Habakkuk	1

Do you wonder why there are so many Biblical references to God as Creator? The reason is clear. The four living creatures and the twenty-four elders in heaven proclaimed the true reason for this important teaching. *Each of the four living creatures had six wings and was covered with eyes all around, even under his wings. Day and night they never stop saying: "Holy, holy, holy is the Lord God Almighty, who was, and is, and is to come." Whenever the living creatures give glory, honor and thanks to him who sits on the throne and who lives for ever and ever, the twenty-four elders fall down before him who sits on the throne, and worship him who lives for ever and ever. They lay their crowns before the throne and say: "You are worthy, our Lord and God, to receive glory and honor and power, for you created all things, and by your will they were created and have their being."* (Rev 4:8-11) We must do no less.

As we see the beauty and complexity of nature, we too are to lift our praise of the Creator. We are to worship the Creator and not His Creation as many have done in times past and as we will see some are doing this today even in our advanced technological world. We must understand that only *by Him all things consist* (Colossians 1:17, KJV). Even the ungodly are without excuse and cannot help but see God in nature. *The heavens declare the glory of God; the skies proclaim the work of his hands. Day after day they pour forth speech; night after night they display knowledge. There is no speech or language where their voice is not heard* (Ps 19:1-3). This important fundamental truth is repeated again for emphasis. *The heavens proclaim his righteousness, and all the peoples see his glory* (Ps 97:6). Those rejecting the God of Creation are without excuse. God supernaturally created the world and all that is in it as it clearly says in the Word. *All* people see His glory in Creation. The Creation of the universe from nothing was the ultimate miracle. It is not logical. Physics cannot explain it. Creation was *not* a slow, gradual process as many today believe. At this point, the Bible could not be clearer (Ps 33:6, 8-9). Creation was clearly *not* a gradual process or the result of mutational errors, but was instantaneous and purposeful. Let us now read and consider carefully what God has done.

Following each book of the Bible is the total number of verses from that particular book. In some cases, a few verses were included to add clarity, but most refer directly to various aspects of creation. In parentheses following each chapter are the number of verses found in that chapter. As you consider these passages, allow them to draw you ever closer to the God of Creation and of our Salvation. Read them in more than one translation. Consider each passage prayerfully before moving on to the next. Allow them to go deep into your spirit, for they are powerful. Allow them to change your worldview. Allow them to bring you to the very throne of grace as you worship the Creator of all things as do the four living creatures and 24 elders. That is the reason so many passages about Creation were given to us. To Him alone is due all honor and glory and praise.

Genesis Total: 58 Verses

Genesis 1, 31 verses
1 In the beginning God created the heavens and the earth.
2 Now the earth was formless and empty, darkness was over the surface of the deep, and the Spirit of God was hovering over the waters.
3 And God said, "Let there be light," and there was light.
4 God saw that the light was good, and he separated the light from the darkness.
5 God called the light "day," and the darkness he called "night." And there was evening, and there was morning-- the first day.
6 And God said, "Let there be an expanse between the waters to separate water from water."
7 So God made the expanse and separated the water under the expanse from the water above it. And it was so.
8 God called the expanse "sky." And there was evening, and there was morning-- the second day.
9 And God said, "Let the water under the sky be gathered to one place, and let dry ground appear." And it was so.
10 God called the dry ground "land," and the gathered waters he called "seas." And God saw that it was good.
11 Then God said, "Let the land produce vegetation: seed-bearing plants and trees on the land that bear fruit with seed in it, according to their various kinds." And it was so.
12 The land produced vegetation: plants bearing seed according to their kinds and trees bearing fruit with seed in it according to their kinds. And God saw that it was good.

13 And there was evening, and there was morning-- the third day.

14 And God said, "Let there be lights in the expanse of the sky to separate the day from the night, and let them serve as signs to mark seasons and days and years,

15 and let them be lights in the expanse of the sky to give light on the earth." And it was so.

16 God made two great lights-- the greater light to govern the day and the lesser light to govern the night. He also made the stars.

17 God set them in the expanse of the sky to give light on the earth,

18 to govern the day and the night, and to separate light from darkness. And God saw that it was good.

19 And there was evening, and there was morning-- the fourth day.

20 And God said, "Let the water teem with living creatures, and let birds fly above the earth across the expanse of the sky."

21 So God created the great creatures of the sea and every living and moving thing with which the water teems, according to their kinds, and every winged bird according to its kind. And God saw that it was good.

22 God blessed them and said, "Be fruitful and increase in number and fill the water in the seas, and let the birds increase on the earth."

23 And there was evening, and there was morning-- the fifth day.

24 And God said, "Let the land produce living creatures according to their kinds: livestock, creatures that move along the ground, and wild animals, each according to its kind." And it was so.

25 God made the wild animals according to their kinds, the livestock according to their kinds, and all the creatures that move along the ground according to their kinds. And God saw that it was good.

26 Then God said, "Let us make man in our image, in our likeness, and let them rule over the fish of the sea and the birds of the air, over the livestock, over all the earth, and over all the creatures that move along the ground."

27 So God created man in his own image, in the image of God he created him; male and female he created them.

28 God blessed them and said to them, "Be fruitful and increase in number; fill the earth and subdue it. Rule over the fish of the sea and the birds of the air and over every living creature that moves on the ground."

29 Then God said, "I give you every seed-bearing plant on the face of the whole earth and every tree that has fruit with seed in it. They will be yours for food.

30 And to all the beasts of the earth and all the birds of the air and all the creatures that move on the ground-- everything that has the breath of life in it-- I give every green plant for food." And it was so.

31 God saw all that he had made, and it was very good. And there was evening, and there was morning-- the sixth day.

Genesis 2, 13 verses

1 Thus the heavens and the earth were completed in all their vast array.

2 By the seventh day God had finished the work he had been doing; so on the seventh day he rested from all his work.

3 And God blessed the seventh day and made it holy, because on it he rested from all the work of creating that he had done.

4 This is the account of the heavens and the earth when they were created. When the LORD God made the earth and the heavens--

7 the LORD God formed the man from the dust of the ground and breathed into his nostrils the breath of life, and the man became a living being.

8 Now the LORD God had planted a garden in the east, in Eden; and there he put the man he had formed.

9 And the LORD God made all kinds of trees grow out of the ground-- trees that were pleasing to the eye and good for food. In the middle of the garden were the tree of life and the tree of the knowledge of good and evil.

18 The LORD God said, "It is not good for the man to be alone. I will make a helper suitable for him."

19 Now the LORD God had formed out of the ground all the beasts of the field and all the birds of the air. He brought them to the man to see what he would name them; and whatever the man called each living creature, that was its name.

20 So the man gave names to all the livestock, the birds of the air and all the beasts of the field. But for Adam no suitable helper was found.

21 So the LORD God caused the man to fall into a deep sleep; and while he was sleeping, he took one of the man's ribs and closed up the place with flesh.

22 Then the LORD God made a woman from the rib he had taken out of the man, and he brought her to the man.

23 The man said, "This is now bone of my bones and flesh of my flesh; she shall be called 'woman,' for she was taken out of man."

Genesis 3, 2 verses

1 Now the serpent was more crafty than any of the wild animals the LORD God had made. He said to the woman, "Did God really say, 'You must not eat from any tree in the garden'?"

19 By the sweat of your brow you will eat your food until you return to the ground, since from it you were taken; for dust you are and to dust you will return."

Genesis 5, 2 verses

1 This is the written account of Adam's line. When God created man, he made him in the likeness of God.

2 He created them male and female and blessed them. And when they were created, he called them "man."

Genesis 6, 2 verses
6 The LORD was grieved that he had made man on the earth, and his heart was filled with pain.
7 So the LORD said, "I will wipe mankind, whom I have created, from the face of the earth-- men and animals, and creatures that move along the ground, and birds of the air-- for I am grieved that I have made them."

Genesis 7, 1 verse
4 Seven days from now I will send rain on the earth for forty days and forty nights, and I will wipe from the face of the earth every living creature I have made."

Genesis 9, 5 verses
6 "Whoever sheds the blood of man, by man shall his blood be shed; for in the image of God has God made man.
13 I have set my rainbow in the clouds, and it will be the sign of the covenant between me and the earth.
14 Whenever I bring clouds over the earth and the rainbow appears in the clouds,
16 Whenever the rainbow appears in the clouds, I will see it and remember the everlasting covenant between God and all living creatures of every kind on the earth."
17 So God said to Noah, "This is the sign of the covenant I have established between me and all life on the earth."

Genesis 14, 2 verses
19 and he blessed Abram, saying, "Blessed be Abram by God Most High, Creator of heaven and earth.
22 But Abram said to the king of Sodom, "I have raised my hand to the LORD, God Most High, Creator of heaven and earth, and have taken an oath.

Exodus Total: 3 Verses

Exodus 4, 1 verse
11 And the LORD said unto him, Who hath made man's mouth? or who maketh the dumb, or deaf, or the seeing, or the blind? have not I the LORD? (KJV)

Exodus 20, 1 verse

11 For in six days the LORD made the heavens and the earth, the sea, and all that is in them, but he rested on the seventh day. Therefore the LORD blessed the Sabbath day and made it holy.

Exodus 31, 1 verse
17 It will be a sign between me and the Israelites forever, for in six days the LORD made the heavens and the earth, and on the seventh day he abstained from work and rested.

Deuteronomy Total: 3 Verses

Deuteronomy 4, 2 verses
18 And when you look up to the sky and see the sun, the moon and the stars-- all the heavenly array-- do not be enticed into bowing down to them and worshiping things the LORD your God has apportioned to all the nations under heaven.
32 Ask now about the former days, long before your time, from the day God created man on the earth; ask from one end of the heavens to the other. Has anything so great as this ever happened, or has anything like it ever been.

Deuteronomy 32, 1 verse
6 Is this the way you repay the LORD, O foolish and unwise people? Is he not your Father, your Creator, who made you and formed you?

II Kings Total: 1 Verse

II Kings 19, 1 verse
15 And Hezekiah prayed before the LORD, and said, O LORD God of Israel, which dwellest between the cherubims, thou art the God, even thou alone, of all the kingdoms of the earth; thou hast made heaven and earth. (KJV)

I Chronicles Total: 2 Verses

I Chronicles 16, 2 verses
25 For great is the LORD and most worthy of praise; he is to be feared above all gods.
26 For all the gods of the nations are idols, but the LORD made the heavens.

II Chronicles Total: 1 Verse

II Chronicles 2 (1 verse)

12 And Hiram added: "Praise be to the LORD, the God of Israel, who made heaven and earth! He has given King David a wise son, endowed with intelligence and discernment, who will build a temple for the LORD and a palace for himself.

Nehemiah Total: 1 Verse

Nehemiah 9, 1 verse
6 You alone are the LORD. You made the heavens, even the highest heavens, and all their starry host, the earth and all that is on it, the seas and all that is in them. You give life to everything, and the multitudes of heaven worship you.

Job Total: 179 Verses

Job 4, 1 verse
17 'Can a mortal be more righteous than God? Can a man be more pure than his Maker?

Job 9, 5 verses
5 He moves mountains without their knowing it and overturns them in his anger.
6 He shakes the earth from its place and makes its pillars tremble.
7 He speaks to the sun and it does not shine; he seals off the light of the stars.
8 He alone stretches out the heavens and treads on the waves of the sea.
9 He is the Maker of the Bear and Orion, the Pleiades and the constellations of the south.

Job 10, 5 verses
8 "Your hands shaped me and made me. Will you now turn and destroy me?
9 Remember that you molded me like clay. Will you now turn me to dust again?
10 Did you not pour me out like milk and curdle me like cheese,
11 clothe me with skin and flesh and knit me together with bones and sinews?
12 You gave me life and showed me kindness, and in your providence watched over my spirit.

Job 12, 4 verses
7 "But ask the animals, and they will teach you, or the birds of the air, and they will tell you;
8 or speak to the earth, and it will teach you, or let the fish of the sea inform you.
9 Which of all these does not know that the hand of the LORD has done this?
10 In his hand is the life of every creature and the breath of all mankind.

Job 20, 1 verse

4 "Surely you know how it has been from of old, ever since man was placed on the earth,

Job 25, 1 verse
2 "Dominion and awe belong to God; he establishes order in the heights of heaven.

Job 26, 8 verses
7 He spreads out the northern [skies] over empty space; he suspends the earth over nothing.
8 He wraps up the waters in his clouds, yet the clouds do not burst under their weight.
9 He covers the face of the full moon, spreading his clouds over it.
10 He marks out the horizon on the face of the waters for a boundary between light and darkness.
11 The pillars of the heavens quake, aghast at his rebuke.
12 By his power he churned up the sea; by his wisdom he cut Rahab to pieces.
13 By his breath the skies became fair; his hand pierced the gliding serpent.
14 And these are but the outer fringe of his works; how faint the whisper we hear of him! Who then can understand the thunder of his power?"

Job 27, 1 verse
3 as long as I have life within me, the breath of God in my nostrils,

Job 28, 5 verses
23 God understands the way to it and he alone knows where it dwells,
24 for he views the ends of the earth and sees everything under the heavens.
25 When he established the force of the wind and measured out the waters,
26 when he made a decree for the rain and a path for the thunderstorm,
27 then he looked at wisdom and appraised it; he confirmed it and tested it.

Job 31, 1 verse
15 Did not he who made me in the womb make them? Did not the same one form us both within our mothers?

Job 32, 1 verse
22 for if I were skilled in flattery, my Maker would soon take me away.

Job 33, 1 verse
4 The Spirit of God has made me; the breath of the Almighty gives me life.

Job 33, 1 verse

6 I am just like you before God; I too have been taken from clay.

Job 34, 4 verses

13 Who appointed him over the earth? Who put him in charge of the whole world?
14 If it were his intention and he withdrew his spirit and breath,
15 all mankind would perish together and man would return to the dust.
19 who shows no partiality to princes and does not favor the rich over the poor, for they are all the work of his hands?

Job 35, 2 verses

10 But no one says, 'Where is God my Maker, who gives songs in the night,
11 who teaches more to us than to the beasts of the earth and makes us wiser than the birds of the air?'

Job 36, 9 verses

3 I get my knowledge from afar; I will ascribe justice to my Maker.
26 How great is God-- beyond our understanding! The number of his years is past finding out.
27 "He draws up the drops of water, which distill as rain to the streams;
28 the clouds pour down their moisture and abundant showers fall on mankind.
29 Who can understand how he spreads out the clouds, how he thunders from his pavilion?
30 See how he scatters his lightning about him, bathing the depths of the sea.
31 This is the way he governs the nations and provides food in abundance.
32 He fills his hands with lightning and commands it to strike its mark.
33 His thunder announces the coming storm; even the cattle make known its approach.

Job 37, 17 verses

2 Listen! Listen to the roar of his voice, to the rumbling that comes from his mouth.
3 He unleashes his lightning beneath the whole heaven and sends it to the ends of the earth.
4 After that comes the sound of his roar; he thunders with his majestic voice. When his voice resounds, he holds nothing back.
5 God's voice thunders in marvelous ways; he does great things beyond our understanding.
6 He says to the snow, 'Fall on the earth,' and to the rain shower, 'Be a mighty downpour.'

7 So that all men he has made may know his work, he stops every man from his labor.

8 The animals take cover; they remain in their dens.

9 The tempest comes out from its chamber, the cold from the driving winds.

10 The breath of God produces ice, and the broad waters become frozen.

11 He loads the clouds with moisture; he scatters his lightning through them.

12 At his direction they swirl around over the face of the whole earth to do whatever he commands them.

13 He brings the clouds to punish men, or to water his earth and show his love.

14 "Listen to this, Job; stop and consider God's wonders.

15 Do you know how God controls the clouds and makes his lightning flash?

16 Do you know how the clouds hang poised, those wonders of him who is perfect in knowledge?

17 You who swelter in your clothes when the land lies hushed under the south wind,

18 can you join him in spreading out the skies, hard as a mirror of cast bronze?

Job 38, 38 verses

4 "Where were you when I laid the earth's foundation? Tell me, if you understand.

5 Who marked off its dimensions? Surely you know! Who stretched a measuring line across it?

6 On what were its footings set, or who laid its cornerstone--

7 while the morning stars sang together and all the angels shouted for joy?

8 "Who shut up the sea behind doors when it burst forth from the womb,

9 when I made the clouds its garment and wrapped it in thick darkness,

10 when I fixed limits for it and set its doors and bars in place,

11 when I said, 'This far you may come and no farther; here is where your proud waves halt'?

12 "Have you ever given orders to the morning, or shown the dawn its place,

13 that it might take the earth by the edges and shake the wicked out of it?

14 The earth takes shape like clay under a seal; its features stand out like those of a garment.

15 The wicked are denied their light, and their upraised arm is broken.

16 "Have you journeyed to the springs of the sea or walked in the recesses of the deep?

17 Have the gates of death been shown to you? Have you seen the gates of the shadow of death?

18 Have you comprehended the vast expanses of the earth? Tell me, if you know all this.

19 "What is the way to the abode of light? And where does darkness reside?

20 Can you take them to their places? Do you know the paths to their dwellings?

21 Surely you know, for you were already born! You have lived so many years!

22 "Have you entered the storehouses of the snow or seen the storehouses of the hail,

23 which I reserve for times of trouble, for days of war and battle?

24 What is the way to the place where the lightning is dispersed, or the place where the east winds are scattered over the earth?

25 Who cuts a channel for the torrents of rain, and a path for the thunderstorm,

26 to water a land where no man lives, a desert with no one in it,

27 to satisfy a desolate wasteland and make it sprout with grass?

28 Does the rain have a father? Who fathers the drops of dew?

29 From whose womb comes the ice? Who gives birth to the frost from the heavens

30 when the waters become hard as stone, when the surface of the deep is frozen?

31 "Can you bind the beautiful Pleiades? Can you loose the cords of Orion?

32 Can you bring forth the constellations in their seasons or lead out the Bear with its cubs?

33 Do you know the laws of the heavens? Can you set up [God's] dominion over the earth?

34 "Can you raise your voice to the clouds and cover yourself with a flood of water?

35 Do you send the lightning bolts on their way? Do they report to you, 'Here we are'?

36 Who endowed the heart with wisdom or gave understanding to the mind?

37 Who has the wisdom to count the clouds? Who can tip over the water jars of the heavens

38 when the dust becomes hard and the clods of earth stick together?

39 "Do you hunt the prey for the lioness and satisfy the hunger of the lions

40 when they crouch in their dens or lie in wait in a thicket?

41 Who provides food for the raven when it's young cry out to God and wander about for lack of food?

Job 39, 30 verses

1 "Do you know when the mountain goats give birth? Do you watch when the doe bears her fawn?

2 Do you count the months till they bear? Do you know the time they give birth?

3 They crouch down and bring forth their young; their labor pains are ended.

4 Their young thrive and grow strong in the wilds; they leave and do not return.

5 "Who let the wild donkey go free? Who untied his ropes?

6 I gave him the wasteland as his home, the salt flats as his habitat.

7 He laughs at the commotion in the town; he does not hear a driver's shout.

8 He ranges the hills for his pasture and searches for any green thing.

9 "Will the wild ox consent to serve you? Will he stay by your manger at night?

10 Can you hold him to the furrow with a harness? Will he till the valleys behind you?

11 Will you rely on him for his great strength? Will you leave your heavy work to him?

12 Can you trust him to bring in your grain and gather it to your threshing floor?

13 "The wings of the ostrich flap joyfully, but they cannot compare with the pinions and feathers of the stork.

14 She lays her eggs on the ground and lets them warm in the sand,

15 unmindful that a foot may crush them, that some wild animal may trample them.

16 She treats her young harshly, as if they were not hers; she cares not that her labor was in vain,

17 for God did not endow her with wisdom or give her a share of good sense.

18 Yet when she spreads her feathers to run, she laughs at horse and rider.

19 "Do you give the horse his strength or clothe his neck with a flowing mane?

20 Do you make him leap like a locust, striking terror with his proud snorting?

21 He paws fiercely, rejoicing in his strength, and charges into the fray.

22 He laughs at fear, afraid of nothing; he does not shy away from the sword.

23 The quiver rattles against his side, along with the flashing spear and lance.

24 In frenzied excitement he eats up the ground; he cannot stand still when the trumpet sounds.

25 At the blast of the trumpet he snorts, 'Aha!' He catches the scent of battle from afar, the shout of commanders and the battle cry.

26 "Does the hawk take flight by your wisdom and spread his wings toward the south?

27 Does the eagle soar at your command and build his nest on high?

28 He dwells on a cliff and stays there at night; a rocky crag is his stronghold.

29 From there he seeks out his food; his eyes detect it from afar.

30 His young ones feast on blood, and where the slain are, there is he."

Job 40, 10 verses

15 "Look at the behemoth, which I made along with you and which feeds on grass like an ox.

16 What strength he has in his loins, what power in the muscles of his belly!

17 His tail sways like a cedar; the sinews of his thighs are close-knit.

18 His bones are tubes of bronze, his limbs like rods of iron.

19 He ranks first among the works of God, yet his Maker can approach him with his sword.

20 The hills bring him their produce, and all the wild animals play nearby.

21 Under the lotus plants he lies, hidden among the reeds in the marsh.

22 The lotuses conceal him in their shadow; the poplars by the stream surround him.

23 When the river rages, he is not alarmed; he is secure, though the Jordan should surge against his mouth.

24 Can anyone capture him by the eyes, or trap him and pierce his nose?

Job 41, 34 verses

1 "Can you pull in the leviathan with a fishhook or tie down his tongue with a rope?

2 Can you put a cord through his nose or pierce his jaw with a hook?

3 Will he keep begging you for mercy? Will he speak to you with gentle words?

4 Will he make an agreement with you for you to take him as your slave for life?

5 Can you make a pet of him like a bird or put him on a leash for your girls?

6 Will traders barter for him? Will they divide him up among the merchants?

7 Can you fill his hide with harpoons or his head with fishing spears?

8 If you lay a hand on him, you will remember the struggle and never do it again!

9 Any hope of subduing him is false; the mere sight of him is overpowering.

10 No one is fierce enough to rouse him. Who then is able to stand against me?

11 Who has a claim against me that I must pay? Everything under heaven belongs to me.

12 "I will not fail to speak of his limbs, his strength and his graceful form.

13 Who can strip off his outer coat? Who would approach him with a bridle?

14 Who dares open the doors of his mouth, ringed about with his fearsome teeth?

15 His back has rows of shields tightly sealed together;

16 each is so close to the next that no air can pass between.

17 They are joined fast to one another; they cling together and cannot be parted.

18 His snorting throws out flashes of light; his eyes are like the rays of dawn.

19 Firebrands stream from his mouth; sparks of fire shoot out.

20 Smoke pours from his nostrils as from a boiling pot over a fire of reeds.

21 His breath sets coals ablaze, and flames dart from his mouth.

22 Strength resides in his neck; dismay goes before him.

23 The folds of his flesh are tightly joined; they are firm and immovable.

24 His chest is hard as rock, hard as a lower millstone.

25 When he rises up, the mighty are terrified; they retreat before his thrashing.

26 The sword that reaches him has no effect, nor does the spear or the dart or the javelin.

27 Iron he treats like straw and bronze like rotten wood.

28 Arrows do not make him flee; sling stones are like chaff to him.

29 A club seems to him but a piece of straw; he laughs at the rattling of the lance.

30 His undersides are jagged potsherds, leaving a trail in the mud like a threshing sledge.

31 He makes the depths churn like a boiling caldron and stirs up the sea like a pot of ointment.

32 Behind him he leaves a glistening wake; one would think the deep had white hair.

33 Nothing on earth is his equal-- a creature without fear.

34 He looks down on all that are haughty; he is king over all that are proud."

Psalms Total: 99 Verses

Psalms 8, 9 verses

1 O LORD, our Lord, how majestic is your name in all the earth! You have set your glory above the heavens.

2 From the lips of children and infants you have ordained praise because of your enemies, to silence the foe and the avenger.

3 When I consider your heavens, the work of your fingers, the moon and the stars, which you have set in place,

4 what is man that you are mindful of him, the son of man that you care for him?

5 You made him a little lower than the heavenly beings and crowned him with glory and honor.

6 You made him ruler over the works of your hands; you put everything under his feet:

7 all flocks and herds, and the beasts of the field,

8 the birds of the air, and the fish of the sea, all that swim the paths of the seas.

9 O LORD, our Lord, how majestic is your name in all the earth!

Psalms 19, 1 verse

1 The heavens declare the glory of God; the skies proclaim the work of his hands.

Psalms 24, 2 verses

1 The earth is the LORD's, and everything in it, the world, and all who live in it;

2 for he founded it upon the seas and established it upon the waters.

Psalms 33, 4 verses

6 By the word of the LORD were the heavens made, their starry host by the breath of his mouth.

7 He gathers the waters of the sea into jars; he puts the deep into storehouses.

8 Let all the earth fear the LORD; let all the people of the world revere him.

9 For he spoke, and it came to be; he commanded, and it stood firm.

Psalms 50, 1 verse

1 The Mighty One, God, the LORD, speaks and summons the earth from the rising of the sun to the place where it sets.

Psalms 65, 2 verses
5 You answer us with awesome deeds of righteousness, O God our Savior, the hope of all the ends of the earth and of the farthest seas,
6 who formed the mountains by your power, having armed yourself with strength,

Psalms 74, 2 verses
16 The day is yours, and yours also the night; you established the sun and moon.
17 It was you who set all the boundaries of the earth; you made both summer and winter.

Psalms 89, 3 verses
4 'I will establish your line forever and make your throne firm through all generations.'" Selah
5 The heavens praise your wonders, O LORD, your faithfulness too, in the assembly of the holy ones.
6 For who in the skies above can compare with the LORD? Who is like the LORD among the heavenly beings?

Psalms 89, 4 verses
9 You rule over the surging sea; when its waves mount up, you still them.
10 You crushed Rahab like one of the slain; with your strong arm you scattered your enemies.
11 The heavens are yours, and yours also the earth; you founded the world and all that is in it.
12 You created the north and the south; Tabor and Hermon sing for joy at your name.

Psalms 90, 1 verse
2 Before the mountains were born or you brought forth the earth and the world, from everlasting to everlasting you are God.

Psalms 95, 4 verses
3 For the LORD is the great God, the great King above all gods.
4 In his hand are the depths of the earth, and the mountain peaks belong to him.
5 The sea is his, for he made it, and his hands formed the dry land.
6 Come, let us bow down in worship, let us kneel before the LORD our Maker;

Psalms 96, 5 verses

1 Sing to the LORD a new song; sing to the LORD, all the earth.

2 Sing to the LORD, praise his name; proclaim his salvation day after day.

3 Declare his glory among the nations, his marvelous deeds among all peoples.

4 For great is the LORD and most worthy of praise; he is to be feared above all gods.

5 For all the gods of the nations are idols, but the LORD made the heavens.

Psalms 100, 1 verse

3 Know that the LORD is God. It is he who made us, and we are his; we are his people, the sheep of his pasture.

Psalms 102, 1 verse

25 In the beginning you laid the foundations of the earth, and the heavens are the work of your hands.

Psalms 103, 2 verses

13 As a father has compassion on his children, so the LORD has compassion on those who fear him;

14 for he knows how we are formed, he remembers that we are dust.

Psalms 104, 9 verses

1 Praise the LORD, O my soul. O LORD my God, you are very great; you are clothed with splendor and majesty.

2 He wraps himself in light as with a garment; he stretches out the heavens like a tent

3 and lays the beams of his upper chambers on their waters. He makes the clouds his chariot and rides on the wings of the wind.

4 He makes winds his messengers, flames of fire his servants.

5 He set the earth on its foundations; it can never be moved.

6 You covered it with the deep as with a garment; the waters stood above the mountains.

7 But at your rebuke the waters fled, at the sound of your thunder they took to flight;

8 they flowed over the mountains, they went down into the valleys, to the place you assigned for them.

9 You set a boundary they cannot cross; never again will they cover the earth.

Psalms 115, 2 verses

15 May you be blessed by the LORD, the Maker of heaven and earth.

16 The highest heavens belong to the LORD, but the earth he has given to man.

Psalms 118, 1 verse
24 This is the day the LORD has made; let us rejoice and be glad in it.

Psalms 119, 1 verse
90 Your faithfulness continues through all generations; you established the earth, and it endures.

Psalms 121, 1 verse
2 My help comes from the LORD, the Maker of heaven and earth.

Psalms 124, 1 verse
8 Our help is in the name of the LORD, the Maker of heaven and earth.

Psalms 134, 1 verse
3 May the LORD, the Maker of heaven and earth, bless you from Zion.

Psalms 135, 1 verse
7 He makes clouds rise from the ends of the earth; he sends lightning with the rain and brings out the wind from his storehouses.

Psalms 136, 7 verses
5 who by his understanding made the heavens, His love endures forever.
6 who spread out the earth upon the waters, His love endures forever.
7 who made the great lights-- His love endures forever.
8 the sun to govern the day, His love endures forever.
9 the moon and stars to govern the night; His love endures forever.
25 and who gives food to every creature. His love endures forever.
26 Give thanks to the God of heaven. His love endures forever.

Psalms 138, 1 verse
8 The LORD will fulfill [his purpose] for me; your love, O LORD, endures forever-- do not abandon the works of your hands.

Psalms 139, 4 verses
13 For you created my inmost being; you knit me together in my mother's womb.
14 I praise you because I am fearfully and wonderfully made; your works are wonderful, I know that full well.
15 My frame was not hidden from you when I was made in the secret place. When I was woven together in the depths of the earth,
16 your eyes saw my unformed body. All the days ordained for me were written in your book before one of them came to be.

Psalms 143, 1 verse

5 I remember the days of long ago; I meditate on all your works and consider what your hands have done.

Psalms 145, 1 verse

10 All you have made will praise you, O LORD; your saints will extol you.

Psalms 145, 2 verses

15 The eyes of all look to you, and you give them their food at the proper time.
16 You open your hand and satisfy the desires of every living thing.

Psalms 146, 2 verses

5 Blessed is he whose help is the God of Jacob, whose hope is in the LORD his God,
6 the Maker of heaven and earth, the sea, and everything in them-- the LORD, who remains faithful forever.

Psalms 147, 7 verses

4 He determines the number of the stars and calls them each by name.
8 He covers the sky with clouds; he supplies the earth with rain and makes grass grow on the hills.
9 He provides food for the cattle and for the young ravens when they call.
15 He sends his command to the earth; his word runs swiftly.
16 He spreads the snow like wool and scatters the frost like ashes.
17 He hurls down his hail like pebbles. Who can withstand his icy blast?
18 He sends his word and melts them; he stirs up his breezes, and the waters flow.

Psalms 148, 13 verses

1 Praise the LORD. Praise the LORD from the heavens, praise him in the heights above.
2 Praise him, all his angels, praise him, all his heavenly hosts.
3 Praise him, sun and moon, praise him, all you shining stars.
4 Praise him, you highest heavens and you waters above the skies.
5 Let them praise the name of the LORD, for he commanded and they were created.
6 He set them in place forever and ever; he gave a decree that will never pass away.
7 Praise the LORD from the earth, you great sea creatures and all ocean depths,
8 lightning and hail, snow and clouds, stormy winds that do his bidding,
9 you mountains and all hills, fruit trees and all cedars,
10 wild animals and all cattle, small creatures and flying birds,

11 kings of the earth and all nations, you princes and all rulers on earth,

12 young men and maidens, old men and children.

13 Let them praise the name of the LORD, for his name alone is exalted; his splendor is above the earth and the heavens.

Psalms 149, 1 verse

2 Let Israel rejoice in their Maker; let the people of Zion be glad in their King.

Proverbs Total: 17 Verses

Proverbs 3, 1 verse

19 By wisdom the LORD laid the earth's foundations, by understanding he set the heavens in place;

Proverbs 8, 10 verses

22 "The LORD brought me forth as the first of his works, before his deeds of old;

23 I was appointed from eternity, from the beginning, before the world began.

24 When there were no oceans, I was given birth, when there were no springs abounding with water;

25 before the mountains were settled in place, before the hills, I was given birth,

26 before he made the earth or its fields or any of the dust of the world.

27 I was there when he set the heavens in place, when he marked out the horizon on the face of the deep,

28 when he established the clouds above and fixed securely the fountains of the deep,

29 when he gave the sea its boundary so the waters would not overstep his command, and when he marked out the foundations of the earth.

30 Then I was the craftsman at his side. I was filled with delight day after day, rejoicing always in his presence,

31 rejoicing in his whole world and delighting in mankind.

Proverbs 14, 1 verse

31 He who oppresses the poor shows contempt for their Maker, but whoever is kind to the needy honors God.

Proverbs 17, 1 verse

5 He who mocks the poor shows contempt for their Maker; whoever gloats over disaster will not go unpunished.

Proverbs 20, 1 verse

12 Ears that hear and eyes that see-- the LORD has made them both.

Proverbs 22, 1 verse
2 Rich and poor have this in common: The LORD is the Maker of them all.

Proverbs 29, 1 verse
13 The poor man and the oppressor have this in common: The LORD gives sight to the eyes of both.

Proverbs 30, 1 verse
4 Who has gone up to heaven and come down? Who has gathered up the wind in the hollow of his hands? Who has wrapped up the waters in his cloak? Who has established all the ends of the earth? What is his name, and the name of his son? Tell me if you know!

Ecclesiastes Total: 5 Verses

Ecclesiastes 3, 2 verses
11 He has made everything beautiful in its time. He has also set eternity in the hearts of men; yet they cannot fathom what God has done from beginning to end.
14 I know that everything God does will endure forever; nothing can be added to it and nothing taken from it. God does it so that men will revere him.

Ecclesiastes 7, 1 verse
29 This only have I found: God made mankind upright, but men have gone in search of many schemes."
Ecclesiastes 11, 1 verse
5 As you do not know the path of the wind, or how the body is formed in a mother's womb, so you cannot understand the work of God, the Maker of all things.

Ecclesiastes 12, 1 verse
1 Remember your Creator in the days of your youth, before the days of trouble come and the years approach when you will say, "I find no pleasure in them"--

Isaiah Total: 43 Verses

Isaiah 17, 1 verse
7 In that day men will look to their Maker and turn their eyes to the Holy One of Israel.

Isaiah 22, 1 verse

11 You built a reservoir between the two walls for the water of the Old Pool, but you did not look to the One who made it, or have regard for the One who planned it long ago.

Isaiah 27, 1 verse
11 When its twigs are dry, they are broken off and women come and make fires with them. For this is a people without understanding; so their Maker has no compassion on them, and their Creator shows them no favor.

Isaiah 29, 1 verse
16 You turn things upside down, as if the potter were thought to be like the clay! Shall what is formed say to him who formed it, "He did not make me"? Can the pot say of the potter, "He knows nothing"?

Isaiah 34, 7 verses
11 The desert owl and screech owl will possess it; the great owl and the raven will nest there. God will stretch out over Edom the measuring line of chaos and the plumb line of desolation.
12 Her nobles will have nothing there to be called a kingdom, all her princes will vanish away.
13 Thorns will overrun her citadels, nettles and brambles her strongholds. She will become a haunt for jackals, a home for owls.
14 Desert creatures will meet with hyenas, and wild goats will bleat to each other; there the night creatures will also repose and find for themselves places of rest.
15 The owl will nest there and lay eggs, she will hatch them, and care for her young under the shadow of her wings; there also the falcons will gather, each with its mate.
16 Look in the scroll of the LORD and read: None of these will be missing, not one will lack her mate. For it is his mouth that has given the order, and his Spirit will gather them together.
17 He allots their portions; his hand distributes them by measure. They will possess it forever and dwell there from generation to generation.

Isaiah 37, 1 verse
16 "O LORD Almighty, God of Israel, enthroned between the cherubim, you alone are God over all the kingdoms of the earth. You have made heaven and earth.

Isaiah 40, 2 verses
21 Do you not know? Have you not heard? Has it not been told you from the beginning? Have you not understood since the earth was founded?

22 He sits enthroned above the circle of the earth, and its people are like grasshoppers. He stretches out the heavens like a canopy, and spreads them out like a tent to live in.

Isaiah 40, 1 verse
26 Lift your eyes and look to the heavens: Who created all these? He who brings out the starry host one by one, and calls them each by name. Because of his great power and mighty strength, not one of them is missing.

Isaiah 40, 1 verse
28 Do you not know? Have you not heard? The LORD is the everlasting God, the Creator of the ends of the earth. He will not grow tired or weary, and his understanding no one can fathom.

Isaiah 42, 1 verse
5 This is what God the LORD says-- he who created the heavens and stretched them out, who spread out the earth and all that comes out of it, who gives breath to its people, and life to those who walk on it:

Isaiah 43, 1 verse
1 But now, this is what the LORD says-- he who created you, O Jacob, he who formed you, O Israel: "Fear not, for I have redeemed you; I have summoned you by name; you are mine.

Isaiah 44, 1 verse
2 This is what the LORD says-- he who made you, who formed you in the womb, and who will help you: Do not be afraid, O Jacob, my servant, Jeshurun, whom I have chosen.

Isaiah 45, 8 verses
5 I am the LORD, and there is no other; apart from me there is no God. I will strengthen you, though you have not acknowledged me,
6 so that from the rising of the sun to the place of its setting men may know there is none besides me. I am the LORD, and there is no other.
7 I form the light and create darkness, I bring prosperity and create disaster; I, the LORD, do all these things.
8 "You heavens above, rain down righteousness; let the clouds shower it down. Let the earth open wide, let salvation spring up, let righteousness grow with it; I, the LORD, have created it.

9 "Woe to him who quarrels with his Maker, to him who is but a potsherd among the potsherds on the ground. Does the clay say to the potter, 'What are you making?' Does your work say, 'He has no hands'?

10 Woe to him who says to his father, 'What have you begotten?' or to his mother, 'What have you brought to birth?'

11 "This is what the LORD says-- the Holy One of Israel, and its Maker: Concerning things to come, do you question me about my children, or give me orders about the work of my hands?

12 It is I who made the earth and created mankind upon it. My own hands stretched out the heavens; I marshaled their starry hosts.

Isaiah 45, 1 verse

18 For this is what the LORD says-- he who created the heavens, he is God; he who fashioned and made the earth, he founded it; he did not create it to be empty, but formed it to be inhabited-- he says: "I am the LORD, and there is no other.

Isaiah 46, 1 verse

11 From the east I summon a bird of prey; from a far-off land, a man to fulfill my purpose. What I have said, that will I bring about; what I have planned, that will I do.

Isaiah 48, 1 verse

13 My own hand laid the foundations of the earth, and my right hand spread out the heavens; when I summon them, they all stand up together.

Isaiah 49, 2 verses

1 Listen to me, you islands; hear this, you distant nations: Before I was born the LORD called me; from my birth he has made mention of my name.

2 He made my mouth like a sharpened sword, in the shadow of his hand he hid me; he made me into a polished arrow and concealed me in his quiver.

Isaiah 49, 1 verse

5 And now the LORD says-- he who formed me in the womb to be his servant to bring Jacob back to him and gather Israel to himself, for I am honored in the eyes of the LORD and my God has been my strength—

Isaiah 50, 1 verse

3 I clothe the sky with darkness and make sackcloth its covering."

Isaiah 51, 1 verse

13 that you forget the LORD your Maker, who stretched out the heavens and laid the foundations of the earth, that you live in constant terror every day because of the wrath of the oppressor, who is bent on destruction? For where is the wrath of the oppressor?

Isaiah 51, 2 verses
15 For I am the LORD your God, who churns up the sea so that its waves roar-- the LORD Almighty is his name.
16 I have put my words in your mouth and covered you with the shadow of my hand-- I who set the heavens in place, who laid the foundations of the earth, and who say to Zion, 'You are my people.'"

Isaiah 54, 2 verses
5 For your Maker is your husband-- the LORD Almighty is his name-- the Holy One of Israel is your Redeemer; he is called the God of all the earth.
16 "See, it is I who created the blacksmith who fans the coals into flame and forges a weapon fit for its work. And it is I who have created the destroyer to work havoc;

Isaiah 57, 1 verse
16 I will not accuse forever, nor will I always be angry, for then the spirit of man would grow faint before me-- the breath of man that I have created.

Isaiah 64, 1 verse
8 Yet, O LORD, you are our Father. We are the clay, you are the potter; we are all the work of your hand.
Isaiah 66, 2 verses
1 This is what the LORD says: "Heaven is my throne, and the earth is my footstool. Where is the house you will build for me? Where will my resting place be?
2 Has not my hand made all these things, and so they came into being?" declares the LORD. "This is the one I esteem: he who is humble and contrite in spirit, and trembles at my word.

Jeremiah Total: 22 Verses

Jeremiah 1, 1 verse
5 "Before I formed you in the womb I knew you, before you were born I set you apart; I appointed you as a prophet to the nations."

Jeremiah 3, 1 verse

3 Therefore the showers have been withheld, and no spring rains have fallen. Yet you have the brazen look of a prostitute; you refuse to blush with shame.

Jeremiah 5, 1 verse
22 Should you not fear me?" declares the LORD. "Should you not tremble in my presence? I made the sand a boundary for the sea, an everlasting barrier it cannot cross. The waves may roll, but they cannot prevail; they may roar, but they cannot cross it.

Jeremiah 10, 2 verses
12 But God made the earth by his power; he founded the world by his wisdom and stretched out the heavens by his understanding.
13 When he thunders, the waters in the heavens roar; he makes clouds rise from the ends of the earth. He sends lightning with the rain and brings out the wind from his storehouses.

Jeremiah 14, 1 verse
Do any of the worthless idols of the nations bring rain? Do the skies themselves send down showers? No, it is you, O LORD our God. Therefore our hope is in you, for you are the one who does all this.

Jeremiah 27, 1 verse
5 With my great power and outstretched arm I made the earth and its people and the animals that are on it, and I give it to anyone I please.

Jeremiah 31, 3 verses
35 This is what the LORD says, he who appoints the sun to shine by day, who decrees the moon and stars to shine by night, who stirs up the sea so that its waves roar-- the LORD Almighty is his name:
36 "Only if these decrees vanish from my sight," declares the LORD, "will the descendants of Israel ever cease to be a nation before me."
37 This is what the LORD says: "Only if the heavens above can be measured and the foundations of the earth below be searched out will I reject all the descendants of Israel because of all they have done," declares the LORD.

Jeremiah 32, 1 verse
17 "Ah, Sovereign LORD, you have made the heavens and the earth by your great power and outstretched arm. Nothing is too hard for you.

Jeremiah 33, 1 verse

2 "This is what the LORD says, he who made the earth, the LORD who formed it and established it-- the LORD is his name:

Jeremiah 51, 5 verses
15 "He made the earth by his power; he founded the world by his wisdom and stretched out the heavens by his understanding.
16 When he thunders, the waters in the heavens roar; he makes clouds rise from the ends of the earth. He sends lightning with the rain and brings out the wind from his storehouses.

19 He who is the Portion of Jacob is not like these, for he is the Maker of all things, including the tribe of his inheritance-- the LORD Almighty is his name.

29 The land trembles and writhes, for the LORD's purposes against Babylon stand-- to lay waste the land of Babylon so that no one will live there.
36 Therefore, this is what the LORD says: "See, I will defend your cause and avenge you; I will dry up her sea and make her springs dry.

Jeremiah 52, 5 verses
3 It was because of the LORD's anger that all this happened to Jerusalem and Judah, and in the end he thrust them from his presence. Now Zedekiah rebelled against the king of Babylon.
4 So in the ninth year of Zedekiah's reign, on the tenth day of the tenth month, Nebuchadnezzar king of Babylon marched against Jerusalem with his whole army. They camped outside the city and built siege works all around it.
5 The city was kept under siege until the eleventh year of King Zedekiah.
6 By the ninth day of the fourth month the famine in the city had become so severe that there was no food for the people to eat.
7 Then the city wall was broken through, and the whole army fled. They left the city at night through the gate between the two walls near the king's garden, though the Babylonians were surrounding the city. They fled toward the Arabah,

Hosea Total: 1 Verse

Hosea 8, 1 verse
14 Israel has forgotten his Maker and built palaces; Judah has fortified many towns. But I will send fire upon their cities that will consume their fortresses.

Amos Total: 13 Verses

Amos 3, 5 verses

4 Does a lion roar in the thicket when he has no prey? Does he growl in his den when he has caught nothing?

5 Does a bird fall into a trap on the ground where no snare has been set? Does a trap spring up from the earth when there is nothing to catch?

6 When a trumpet sounds in a city, do not the people tremble? When disaster comes to a city, has not the LORD caused it?

7 Surely the Sovereign LORD does nothing without revealing his plan to his servants the prophets.

8 The lion has roared-- who will not fear? The Sovereign LORD has spoken-- who can but prophesy?

Amos 4, 6 verses

7 I also withheld rain from you when the harvest was still three months away. I sent rain on one town, but withheld it from another. One field had rain; another had none and dried up.

9 "Many times I struck your gardens and vineyards, I struck them with blight and mildew. Locusts devoured your fig and olive trees, yet you have not returned to me," declares the LORD.

10 "I sent plagues among you as I did to Egypt. I killed your young men with the sword, along with your captured horses. I filled your nostrils with the stench of your camps, yet you have not returned to me," declares the LORD.

11 "I overthrew some of you as I overthrew Sodom and Gomorrah. You were like a burning stick snatched from the fire, yet you have not returned to me," declares the LORD.

12 "Therefore this is what I will do to you, Israel, and because I will do this to you, prepare to meet your God, O Israel."

13 He who forms the mountains, creates the wind, and reveals his thoughts to man, he who turns dawn to darkness, and treads the high places of the earth-- the LORD God Almighty is his name.

Amos 5, 1 verse

8 (he who made the Pleiades and Orion, who turns blackness into dawn and darkens day into night, who calls for the waters of the sea and pours them out over the face of the land-- the LORD is his name--

Amos 9, 2 verses

5 The Lord, the LORD Almighty, he who touches the earth and it melts, and all who live in it mourn-- the whole land rises like the Nile, then sinks like the river of Egypt--

6 he who builds his lofty palace in the heavens and sets its foundation on the earth, who calls for the waters of the sea and pours them out over the face of the land-- the LORD is his name.

Jonah Total: 4 Verses

Jonah 1, 3 verses
4 Then the LORD sent a great wind on the sea, and such a violent storm arose that the ship threatened to break up.
9 He answered, "I am a Hebrew and I worship the LORD, the God of heaven, who made the sea and the land."
17 But the LORD provided a great fish to swallow Jonah, and Jonah was inside the fish three days and three nights.

Jonah 2, 1 verse
10 And the LORD commanded the fish, and it vomited Jonah onto dry land.

Habakkuk Total: 1 Verse

Habakkuk 3, 1 verse
11 Sun and moon stood still in the heavens at the glint of your flying arrows, at the lightning of your flashing spear.

Zechariah Total: 2 Verses

Zechariah 10, 1 verse
1 Ask the LORD for rain in the springtime; it is the LORD who makes the storm clouds. He gives showers of rain to men, and plants of the field to everyone.

Zechariah 12, 1 verse
1 This is the word of the LORD concerning Israel. The LORD, who stretches out the heavens, who lays the foundation of the earth, and who forms the spirit of man within him, declares:

Malachi Total: 1 Verse

Malachi 2, 1 verse
10 Have we not all one Father? Did not one God create us? Why do we profane the covenant of our fathers by breaking faith with one another?

New Testament Creation References

Through Him all things were made (John 1:3a).

The Old Testament is God's revelation about himself, His character and His chosen people. The New Testament reveals God's Son and the good news of Salvation. It also provides historical information about the early church and prophetic insight into events in these last days before His return and final judgment. Following is a compilation of scripture references about God as Creator found in the New Testament. It is obvious the recognition of God as Creator remains as important in the Church Age as it was in Old Testament times. Sadly, many today no longer accept the account of creation as described in Scripture. For them, naturalistic evolution has replaced supernatural creation. Rejection of the creation account in the Bible weakens confidence in the entire Word of God and can result in drifting farther and farther from Biblical truth. The six days of Creation are taken by many today as allegorical or myth. The miracles of the Bible are doubted and Christianity becomes no different from other world religions. The Biblical account of creation is central to the Christian faith and is needed today in our increasingly secular and anti-Christian world.

The following table provides a summary of Creation verses by books of the New Testament. Hebrews and Acts provide the most references to God as Creator, but there are other important passages throughout the New Testament. Such references cannot be ignored. Consider each verse in context and read from other translations for additional insight. We were warned that in these last days many would come doubting in the Creation and the Flood due to "willful ignorance" (2 Pet. 3:3-7, KJV). That time has arrived. Let us not be ignorant in these important issues. The origins debate is becoming increasingly popular in the media and even on certain television shows. Let's keep the discussion alive with informed debate.

43 Creation verses in the New Testament

Book	verses	Book	verses	Book	verses
Matthew	3	Romans	2	James	1
Mark	2	1 Corinthians	2	2 Peter	1
Luke	2	Colossians	2	Revelation	1
John	4	1 Timothy	3		
Acts	8	Hebrews	12		

Mathew Total: 3 Verses

Matt 6, 1 verse
30 If that is how God clothes the grass of the field, which is here today and tomorrow is thrown into the fire, will he not much more clothe you, O you of little faith?

Matt 13, 1 verse
35 So was fulfilled what was spoken through the prophet: "I will open my mouth in parables, I will utter things hidden since the creation of the world."

Matt 25, 1 verse
34 "Then the King will say to those on his right, 'Come, you who are blessed by my Father; take your inheritance, the kingdom prepared for you since the creation of the world.

Mark Total: 2 Verses

Mark 10, 1 verse
6 But at the beginning of creation God 'made them male and female.'

Mark 13, 1 verse

19 because those will be days of distress unequaled from the beginning, when God created the world, until now-- and never to be equaled again.

Luke Total: 2 Verses

Luke 3, 1 verse

38 the son of Enosh, the son of Seth, the son of Adam, the son of God.

Luke 12 (1 verse)

28 If that is how God clothes the grass of the field, which is here today, and tomorrow is thrown into the fire, how much more will he clothe you, O you of little faith!

John Total: 4 Verses

John 1, 3 verses

1 In the beginning was the Word, and the Word was with God, and the Word was God.
2 He was with God in the beginning.
3 Through him all things were made; without him nothing was made that has been made.

John 17, 1 verse

24 Father, I want those you have given me to be with me where I am, and to see my glory, the glory you have given me because you loved me before the creation of the world.

Acts Total: 8 Verses

Acts 4, 1 verse

24 When they heard this, they raised their voices together in prayer to God. "Sovereign Lord," they said, "you made the heaven and the earth and the sea, and everything in them.

Acts 14, 1 verse

15 "Men, why are you doing this? We too are only men, human like you. We are bringing you good news, telling you to turn from these worthless things to the living God, who made heaven and earth and sea and everything in them.

Acts 17, 6 verses

24 The God who made the world and everything in it is the Lord of heaven and earth and does not live in temples built by hands.

25 And he is not served by human hands, as if he needed anything, because he himself gives all men life and breath and everything else.

26 From one man he made every nation of men, that they should inhabit the whole earth; and he determined the times set for them and the exact places where they should live.

27 God did this so that men would seek him and perhaps reach out for him and find him, though he is not far from each one of us.

28 'For in him we live and move and have our being.' As some of your own poets have said, 'We are his offspring.'

29 "Therefore since we are God's offspring, we should not think that the divine being is like gold or silver or stone-- an image made by man's design and skill.

Romans Total: 2 Verses

Romans 1, 2 verses

20 For since the creation of the world God's invisible qualities-- his eternal power and divine nature-- have been clearly seen, being understood from what has been made, so that men are without excuse.

25 They exchanged the truth of God for a lie, and worshiped and served created things rather than the Creator-- who is forever praised. Amen.

1 Corinthians Total: 4 Verses

1 Corinthians 8, 1 verse

6 yet for us there is but one God, the Father, from whom all things came and for whom we live; and there is but one Lord, Jesus Christ, through whom all things came and through whom we live.

1 Corinthians 11, 1 verse

9 neither was man created for woman, but woman for man.

1 Corinthians 12, 2 verses

18 But in fact God has arranged the parts in the body, every one of them, just as he wanted them to be.

24 while our presentable parts need no special treatment. But God has combined the members of the body and has given greater honor to the parts that lacked it,

2 Corinthians Total: 1 Verse

2 Corinthians 5, 1 verse
5 Now it is God who has made us for this very purpose and has given us the Spirit as a deposit, guaranteeing what is to come.

Colossians Total: 2 Verses

Colossians 1, 2 verses
16 For by him all things were created: things in heaven and on earth, visible and invisible, whether thrones or powers or rulers or authorities; all things were created by him and for him.
17 He is before all things, and in him all things hold together.

1 Timothy Total: 3 Verses

1 Tim 2, 1 verse
13 For Adam was formed first, then Eve.
1 Tim 4, 2 verses
3 They forbid people to marry and order them to abstain from certain foods, which God created to be received with thanksgiving by those who believe and who know the truth.
4 For everything God created is good, and nothing is to be rejected if it is received with thanksgiving,

Hebrews Total: 12 Verses

Hebrews 1, 3 verses
2 but in these last days he has spoken to us by his Son, whom he appointed heir of all things, and through whom he made the universe.
10 He also says, "In the beginning, O Lord, you laid the foundations of the earth, and the heavens are the work of your hands.
11 They will perish, but you remain; they will all wear out like a garment.

Hebrews 2, 5 verses
7 You made him a little lower than the angels; you crowned him with glory and honor
8 and put everything under his feet." In putting everything under him, God left nothing that is not subject to him. Yet at present we do not see everything subject to him.
9 But we see Jesus, who was made a little lower than the angels, now crowned with glory and honor because he suffered death, so that by the grace of God he might taste death for everyone.

10 In bringing many sons to glory, it was fitting that God, for whom and through whom everything exists, should make the author of their salvation perfect through suffering.

11 Both the one who makes men holy and those who are made holy are of the same family. So Jesus is not ashamed to call them brothers.

Hebrews 3, 1 verse
4 For every house is built by someone, but God is the builder of everything.

Hebrews 4, 2 verses
3 Now we who have believed enter that rest, just as God has said, "So I declared on oath in my anger, 'They shall never enter my rest.'" And yet his work has been finished since the creation of the world.

4 For somewhere he has spoken about the seventh day in these words: "And on the seventh day God rested from all his work."

Hebrews 11, 1 verse
3 By faith we understand that the universe was formed at God's command, so that what is seen was not made out of what was visible.

James Total: 1 Verse

James 1, 1 verse
18 He chose to give us birth through the word of truth, that we might be a kind of first fruits of all he created.

I Peter Total: 1 Verse

I Peter 4, 1 verse
19 So then, those who suffer according to God's will should commit themselves to their faithful Creator and continue to do good.

2 Peter Total: 1 Verse

2 Peter 3, 1 verse
5 But they deliberately forget that long ago by God's word the heavens existed and the earth was formed out of water and by water.

Revelation Total: 1 Verse

Revelation 4, 1 verse

11 You are worthy, our Lord and God, to receive glory and honor and power, for you created all things, and by your will they were created and have their being.

Chapter 3
God Remains Active in Creation

He is before all things, and in him all things hold together.
(Col 1:17, NIV)

An atom

The widespread acceptance of evolution has raised doubts in the minds of many people about the relevance and reliability of the Bible. This is particularly true concerning the Biblical account of Creation and the Flood, but spills over into many other areas including the miracles of both the Old and New Testaments. Many Christians, several denominations and even some Bible seminaries no longer accept verbal inerrancy of scripture including especially the Biblical account of creation and the global flood. Some people try desperately to cling to a belief in God and attempt to reconcile Him with the modern secular view of evolution. They attempt to keep one foot in each camp. As a result, they accept some form of theistic evolution or the view that perhaps God started the process, but then stepped back to let nature "run its course." This is neither defensible from scripture nor acceptable in secular science for evolution is always defined as a fully natural and undirected process. If God played any part in the process it is not evolution.

There are also spiritual risks, for such a view no longer sees God as caring about mankind in an ongoing and meaningful way. Many have bought into the evolution mantra that human life is nothing but "matter in motion." Life has no meaning and the Creator is no longer involved with creation or cares about people. Again, such an impersonal view of God is not supported by scripture and those lingering doubts soon take on a life of their own. Unresolved doubts cause one to question the truth and relevance of all scripture. One common result is Bible study no longer seems important and is neglected. Prayer becomes pointless and sporadic. Personal faith is weakened and witnessing to others ceases.

Perhaps the best way to resolve these doubts is to see that God remains active and involved with His Creation today as He has been throughout history. He did not create the universe and walk away, but is intimately involved with every aspect of creation including especially His unwavering love for mankind. This is an important topic, as evidenced by over one thousand scriptural references demonstrating God's continuing involvement in His creation. Any topic mentioned that many times in scripture must be important, yet we seldom hear this from the pulpit.

Not only did God create *all* things in the beginning, but He is actively involved with creation as the ultimate caregiver. In spite of what many people today say, God did *not* just wind up the universe to start it running and step away. His ongoing attention to details of Creation and intervention in human life shows His continuing love for man and concern for all of creation. We must understand that God remains active even today in creation for ***by Him all things consist*** (Colossians 1:17, KJV). Consider this familiar passage: ***Are not two sparrows sold for a penny? Yet not one of them will fall to the ground apart from the will of your Father. And even the very hairs of your head are all numbered. So don't be afraid; you are worth more than many sparrows*** (Matt 10:29-32a, NIV). Notice God is not merely aware of the sparrow's plight, but it was only by his permissive will that they fell to the ground. We are of much greater value to God than a mere sparrow for He loved us enough to give us His son. How can one not be moved by the statement that even the hair of our heads is numbered by our loving, ever watchful and all-knowing God?

We must worship God not only as our Creator, but also as the One who is actively caring for us and for all of Creation. His love and continual watching over us is beyond measure or human comprehension. This view has strong Biblical support represented by over one thousand scripture verses, yet it seems few Christians are aware of its importance or relevance in our increasingly secular and anti-Christian world. As with Creation, any theme repeated that many times

96

in the Bible is important and worthy of our thoughtful contemplation and our gratefulness. Again, the reader is strongly encouraged to think about each passage and read it in more than one translation. We must get this concept deep inside our spirit. It will change the way we view the world and ourselves, for as His image bearers we are of great value to the God of Creation. More importantly, it will change how we view and treat others. The passages are listed in the same format used in the previous chapter dealing with God as Creator. Also it will be obvious that some passages overlap and speak of God as both Creator and active in creation. Such verses appear in both listings. Again, as in the preceding chapter some verses were included for context to in order to make the meaning more clear. No doubt other verses should have been included, but were missed.

The books in the Bible with the most references to God as active in creation are interesting. Most Bible scholars agree Job is the oldest book in the Bible and it has the most references to God as active in creation (174) as it did regarding God as Creator (179). We thus have two important characteristics of God given long before anything else was written about Him. He was seen early on as both Creator and as the all-caring God. That certainly suggests the importance of these two characteristics and are worthy of thoughtful contemplation. God's active caring is also stressed in Hebrews (106) and Colossians (102) showing the importance of this teaching in the New Testament. Psalms is fourth with 75 verses underscoring the importance of God as Sustainer and as our Creator with 99 verses.

1,002 Verses Proclaiming God Remains Active in Creation

Book	verses	Book	verses	Book	verses
Genesis	16	John	57	1 Timothy	2
Exodus	3	Acts	89	2 Timothy	3
Job	174	Romans	80	Titus	3
Psalms	75	1 Corinthians	26	Hebrew	106
Proverbs	11	2 Corinthians	20	James	22
Ecclesiastes	3	Galatians	14	1 Peter	9
Isaiah	33	Ephesians	18	2 Peter	14
Jeremiah	12	Philippians	1	1 John	23
Mathew	17	Colossians	102	2 John	4
Mark	2	1 Thessalonians	7		
Luke	49	2 Thessalonians	7		

Genesis Total: 16 Verses

Genesis 2, 7 verses
3 And God blessed the seventh day and made it holy, because on it he rested from all the work of creating that he had done.
8 Now the LORD God had planted a garden in the east, in Eden; and there he put the man he had formed.
9 And the LORD God made all kinds of trees grow out of the ground-- trees that were pleasing to the eye and good for food. In the middle of the garden were the tree of life and the tree of the knowledge of good and evil.
18 The LORD God said, "It is not good for the man to be alone. I will make a helper suitable for him."
21 So the LORD God caused the man to fall into a deep sleep; and while he was sleeping, he took one of the man's ribs and closed up the place with flesh.
22 Then the LORD God made a woman from the rib he had taken out of the man, and he brought her to the man.
23 The man said, "This is now bone of my bones and flesh of my flesh; she shall be called 'woman,' for she was taken out of man."

Genesis 6, 2 verses
6 The LORD was grieved that he had made man on the earth, and his heart was filled with pain.
7 So the LORD said, "I will wipe mankind, whom I have created, from the face of the earth-- men and animals, and creatures that move along the ground, and birds of the air-- for I am grieved that I have made them."

Genesis 7, 1 verse
4 Seven days from now I will send rain on the earth for forty days and forty nights, and I will wipe from the face of the earth every living creature I have made."

Genesis 9, 5 verses
6 "Whoever sheds the blood of man, by man shall his blood be shed; for in the image of God has God made man.
13 I have set my rainbow in the clouds, and it will be the sign of the covenant between me and the earth.
14 Whenever I bring clouds over the earth and the rainbow appears in the clouds,
16 Whenever the rainbow appears in the clouds, I will see it and remember the everlasting covenant between God and all living creatures of every kind on the earth."

17 So God said to Noah, "This is the sign of the covenant I have established between me and all life on the earth."

Genesis 14, 1 verse
19 and he blessed Abram, saying, "Blessed be Abram by God Most High, Creator of heaven and earth.

Job Total: 174 Verses

Job 9, 5 verses
5 He moves mountains without their knowing it and overturns them in his anger.
6 He shakes the earth from its place and makes its pillars tremble.
7 He speaks to the sun and it does not shine; he seals off the light of the stars.
8 He alone stretches out the heavens and treads on the waves of the sea.
9 He is the Maker of the Bear and Orion, the Pleiades and the constellations of the south.

Job 10, 5 verses
8 "Your hands shaped me and made me. Will you now turn and destroy me?
9 Remember that you molded me like clay. Will you now turn me to dust again?
10 Did you not pour me out like milk and curdle me like cheese,
11 clothe me with skin and flesh and knit me together with bones and sinews?
12 You gave me life and showed me kindness, and in your providence watched over my spirit.

Job 12, 4 verses
7 "But ask the animals, and they will teach you, or the birds of the air, and they will tell you;
8 or speak to the earth, and it will teach you, or let the fish of the sea inform you.
9 Which of all these does not know that the hand of the LORD has done this?
10 In his hand is the life of every creature and the breath of all mankind.

Job 20, 1 verses
4 "Surely you know how it has been from of old, ever since man was placed on the earth,

Job 25, 1 verses
2 "Dominion and awe belong to God; he establishes order in the heights of heaven.

Job 26, 8 verses

7 He spreads out the northern [skies] over empty space; he suspends the earth over nothing.

8 He wraps up the waters in his clouds, yet the clouds do not burst under their weight.

9 He covers the face of the full moon, spreading his clouds over it.

10 He marks out the horizon on the face of the waters for a boundary between light and darkness.

11 The pillars of the heavens quake, aghast at his rebuke.

12 By his power he churned up the sea; by his wisdom he cut Rahab to pieces.

13 By his breath the skies became fair; his hand pierced the gliding serpent.

14 And these are but the outer fringe of his works; how faint the whisper we hear of him! Who then can understand the thunder of his power?"

Job 28, 5 verses

23 God understands the way to it and he alone knows where it dwells,

24 for he views the ends of the earth and sees everything under the heavens.

25 When he established the force of the wind and measured out the waters,

26 when he made a decree for the rain and a path for the thunderstorm,

27 then he looked at wisdom and appraised it; he confirmed it and tested it.

Job 31, 1 verse

15 Did not he who made me in the womb make them? Did not the same one form us both within our mothers?

Job 33, 1 verse

4 The Spirit of God has made me; the breath of the Almighty gives me life.

Job 33, 1 verse

6 I am just like you before God; I too have been taken from clay.

Job 34, 4 verse

13 Who appointed him over the earth? Who put him in charge of the whole world?

14 If it were his intention and he withdrew his spirit and breath,

15 all mankind would perish together and man would return to the dust.

19 who shows no partiality to princes and does not favor the rich over the poor, for they are all the work of his hands?

Job 35, 2 verses

10 But no one says, 'Where is God my Maker, who gives songs in the night,

11 who teaches more to us than to the beasts of the earth and makes us wiser than the birds of the air?'

Job 36, 7 verses

27 "He draws up the drops of water, which distill as rain to the streams;

28 the clouds pour down their moisture and abundant showers fall on mankind.

29 Who can understand how he spreads out the clouds, how he thunders from his pavilion?

30 See how he scatters his lightning about him, bathing the depths of the sea.

31 This is the way he governs the nations and provides food in abundance.

32 He fills his hands with lightning and commands it to strike its mark.

33 His thunder announces the coming storm; even the cattle make known its approach.

Job 37, 17 verses

2 Listen! Listen to the roar of his voice, to the rumbling that comes from his mouth.

3 He unleashes his lightning beneath the whole heaven and sends it to the ends of the earth.

4 After that comes the sound of his roar; he thunders with his majestic voice. When his voice resounds, he holds nothing back.

5 God's voice thunders in marvelous ways; he does great things beyond our understanding.

6 He says to the snow, 'Fall on the earth,' and to the rain shower, 'Be a mighty downpour.'

7 So that all men he has made may know his work, he stops every man from his labor.

8 The animals take cover; they remain in their dens.

9 The tempest comes out from its chamber, the cold from the driving winds.

10 The breath of God produces ice, and the broad waters become frozen.

11 He loads the clouds with moisture; he scatters his lightning through them.

12 At his direction they swirl around over the face of the whole earth to do whatever he commands them.

13 He brings the clouds to punish men, or to water his earth and show his love.

14 "Listen to this, Job; stop and consider God's wonders.

15 Do you know how God controls the clouds and makes his lightning flash?

16 Do you know how the clouds hang poised, those wonders of him who is perfect in knowledge?

17 You who swelter in your clothes when the land lies hushed under the south wind,

18 can you join him in spreading out the skies, hard as a mirror of cast bronze?

Job 38, 38 verses

4 "Where were you when I laid the earth's foundation? Tell me, if you understand.

5 Who marked off its dimensions? Surely you know! Who stretched a measuring line across it?

6 On what were its footings set, or who laid its cornerstone--

7 while the morning stars sang together and all the angels shouted for joy?

8 "Who shut up the sea behind doors when it burst forth from the womb,

9 when I made the clouds its garment and wrapped it in thick darkness,

10 when I fixed limits for it and set its doors and bars in place,

11 when I said, 'This far you may come and no farther; here is where your proud waves halt'?

12 "Have you ever given orders to the morning, or shown the dawn its place,

13 that it might take the earth by the edges and shake the wicked out of it?

14 The earth takes shape like clay under a seal; its features stand out like those of a garment.

15 The wicked are denied their light, and their upraised arm is broken.

16 "Have you journeyed to the springs of the sea or walked in the recesses of the deep?

17 Have the gates of death been shown to you? Have you seen the gates of the shadow of death?

18 Have you comprehended the vast expanses of the earth? Tell me, if you know all this.

19 "What is the way to the abode of light? And where does darkness reside?

20 Can you take them to their places? Do you know the paths to their dwellings?

21 Surely you know, for you were already born! You have lived so many years!

22 "Have you entered the storehouses of the snow or seen the storehouses of the hail,

23 which I reserve for times of trouble, for days of war and battle?

24 What is the way to the place where the lightning is dispersed, or the place where the east winds are scattered over the earth?

25 Who cuts a channel for the torrents of rain, and a path for the thunderstorm,

26 to water a land where no man lives, a desert with no one in it,

27 to satisfy a desolate wasteland and make it sprout with grass?

28 Does the rain have a father? Who fathers the drops of dew?

29 From whose womb comes the ice? Who gives birth to the frost from the heavens

30 when the waters become hard as stone, when the surface of the deep is frozen?

31 "Can you bind the beautiful Pleiades? Can you loose the cords of Orion?

32 Can you bring forth the constellations in their seasons or lead out the Bear with its cubs?

33 Do you know the laws of the heavens? Can you set up [God's] dominion over the earth?

34 "Can you raise your voice to the clouds and cover yourself with a flood of water?

35 Do you send the lightning bolts on their way? Do they report to you, 'Here we are'?

36 Who endowed the heart with wisdom or gave understanding to the mind?

37 Who has the wisdom to count the clouds? Who can tip over the water jars of the heavens

38 when the dust becomes hard and the clods of earth stick together?

39 "Do you hunt the prey for the lioness and satisfy the hunger of the lions

40 when they crouch in their dens or lie in wait in a thicket?

41 Who provides food for the raven when its young cry out to God and wander about for lack of food?

Job 39, 30 verses

1 "Do you know when the mountain goats give birth? Do you watch when the doe bears her fawn?

2 Do you count the months till they bear? Do you know the time they give birth?

3 They crouch down and bring forth their young; their labor pains are ended.

4 Their young thrive and grow strong in the wilds; they leave and do not return.

5 "Who let the wild donkey go free? Who untied his ropes?

6 I gave him the wasteland as his home, the salt flats as his habitat.

7 He laughs at the commotion in the town; he does not hear a driver's shout.

8 He ranges the hills for his pasture and searches for any green thing.

9 "Will the wild ox consent to serve you? Will he stay by your manger at night?

10 Can you hold him to the furrow with a harness? Will he till the valleys behind you?

11 Will you rely on him for his great strength? Will you leave your heavy work to him?

12 Can you trust him to bring in your grain and gather it to your threshing floor?

13 "The wings of the ostrich flap joyfully, but they cannot compare with the pinions and feathers of the stork.

14 She lays her eggs on the ground and lets them warm in the sand,

15 unmindful that a foot may crush them, that some wild animal may trample them.

16 She treats her young harshly, as if they were not hers; she cares not that her labor was in vain,

17 for God did not endow her with wisdom or give her a share of good sense.

18 Yet when she spreads her feathers to run, she laughs at horse and rider.

19 "Do you give the horse his strength or clothe his neck with a flowing mane?

20Do you make him leap like a locust, striking terror with his proud snorting?

21 He paws fiercely, rejoicing in his strength, and charges into the fray.

22 He laughs at fear, afraid of nothing; he does not shy away from the sword.

23 The quiver rattles against his side, along with the flashing spear and lance.

24 In frenzied excitement he eats up the ground; he cannot stand still when the trumpet sounds.

25 At the blast of the trumpet he snorts, 'Aha!' He catches the scent of battle from afar, the shout of commanders and the battle cry.

26 "Does the hawk take flight by your wisdom and spread his wings toward the south?

27 Does the eagle soar at your command and build his nest on high?

28 He dwells on a cliff and stays there at night; a rocky crag is his stronghold.

29 From there he seeks out his food; his eyes detect it from afar.

30 His young ones feast on blood, and where the slain are, there is he."

Job 40, 10 verses

15 "Look at the behemoth, which I made along with you and which feeds on grass like an ox.

16 What strength he has in his loins, what power in the muscles of his belly!

17 His tail sways like a cedar; the sinews of his thighs are close-knit.

18 His bones are tubes of bronze, his limbs like rods of iron.

19 He ranks first among the works of God, yet his Maker can approach him with his sword.

20 The hills bring him their produce, and all the wild animals play nearby.

21 Under the lotus plants he lies, hidden among the reeds in the marsh.

22 The lotuses conceal him in their shadow; the poplars by the stream surround him.

23 When the river rages, he is not alarmed; he is secure, though the Jordan should surge against his mouth.

24 Can anyone capture him by the eyes, or trap him and pierce his nose?

Job 41, 34 verses

1 "Can you pull in the leviathan with a fishhook or tie down his tongue with a rope?

2 Can you put a cord through his nose or pierce his jaw with a hook?

3 Will he keep begging you for mercy? Will he speak to you with gentle words?

4 Will he make an agreement with you for you to take him as your slave for life?

5 Can you make a pet of him like a bird or put him on a leash for your girls?

6 Will traders barter for him? Will they divide him up among the merchants?

7 Can you fill his hide with harpoons or his head with fishing spears?

8 If you lay a hand on him, you will remember the struggle and never do it again!

9 Any hope of subduing him is false; the mere sight of him is overpowering.

10 No one is fierce enough to rouse him. Who then is able to stand against me?

11 Who has a claim against me that I must pay? Everything under heaven belongs to me.

12 "I will not fail to speak of his limbs, his strength and his graceful form.

13 Who can strip off his outer coat? Who would approach him with a bridle?

14 Who dares open the doors of his mouth, ringed about with his fearsome teeth?

15 His back has rows of shields tightly sealed together;

16 each is so close to the next that no air can pass between.

17 They are joined fast to one another; they cling together and cannot be parted.

18 His snorting throws out flashes of light; his eyes are like the rays of dawn.

19 Firebrands stream from his mouth; sparks of fire shoot out.

20 Smoke pours from his nostrils as from a boiling pot over a fire of reeds.

21 His breath sets coals ablaze, and flames dart from his mouth.

22 Strength resides in his neck; dismay goes before him.

23 The folds of his flesh are tightly joined; they are firm and immovable.

24 His chest is hard as rock, hard as a lower millstone.

25 When he rises up, the mighty are terrified; they retreat before his thrashing.

26 The sword that reaches him has no effect, nor does the spear or the dart or the javelin.

27 Iron he treats like straw and bronze like rotten wood.

28 Arrows do not make him flee; sling stones are like chaff to him.

29 A club seems to him but a piece of straw; he laughs at the rattling of the lance.

30 His undersides are jagged potsherds, leaving a trail in the mud like a threshing sledge.

31 He makes the depths churn like a boiling caldron and stirs up the sea like a pot of ointment.

32 Behind him he leaves a glistening wake; one would think the deep had white hair.

33 Nothing on earth is his equal-- a creature without fear.

34 He looks down on all that are haughty; he is king over all that are proud."

Psalms Total: 75 Verses

Psalms 8, 9 verses

1 O LORD, our Lord, how majestic is your name in all the earth! You have set your glory above the heavens.

2 From the lips of children and infants you have ordained praise because of your enemies, to silence the foe and the avenger.

3 When I consider your heavens, the work of your fingers, the moon and the stars, which you have set in place,

4 what is man that you are mindful of him, the son of man that you care for him?

5 You made him a little lower than the heavenly beings and crowned him with glory and honor.

6 You made him ruler over the works of your hands; you put everything under his feet:

7 all flocks and herds, and the beasts of the field,

8 the birds of the air, and the fish of the sea, all that swim the paths of the seas.

9 O LORD, our Lord, how majestic is your name in all the earth!

Psalms 19, 1 verse

1 The heavens declare the glory of God; the skies proclaim the work of his hands.

Psalms 24, 2 verses

1 The earth is the LORD's, and everything in it, the world, and all who live in it;

2 for he founded it upon the seas and established it upon the waters.

Psalms 33, 4 verses

6 By the word of the LORD were the heavens made, their starry host by the breath of his mouth.

7 He gathers the waters of the sea into jars; he puts the deep into storehouses.

8 Let all the earth fear the LORD; let all the people of the world revere him.

9 For he spoke, and it came to be; he commanded, and it stood firm.

Psalms 50, 1 verse

1 The Mighty One, God, the LORD, speaks and summons the earth from the rising of the sun to the place where it sets.

Psalms 65, 2 verses

5 You answer us with awesome deeds of righteousness, O God our Savior, the hope of all the ends of the earth and of the farthest seas,

6 who formed the mountains by your power, having armed yourself with strength,

Psalms 74, 2 verses

16 The day is yours, and yours also the night; you established the sun and moon.

17 It was you who set all the boundaries of the earth; you made both summer and winter.

Psalms 89, 7 verses

4 I will establish your line forever and make your throne firm through all generations.'" Selah

5 The heavens praise your wonders, O LORD, your faithfulness too, in the assembly of the holy ones.

6 For who in the skies above can compare with the LORD? Who is like the LORD among the heavenly beings?

Psalms 89, 4 verses

9 You rule over the surging sea; when its waves mount up, you still them.

10 You crushed Rahab like one of the slain; with your strong arm you scattered your enemies.

11 The heavens are yours, and yours also the earth; you founded the world and all that is in it.

12 You created the north and the south; Tabor and Hermon sing for joy at your name.

Psalms 90, 1 verse

2 Before the mountains were born or you brought forth the earth and the world, from everlasting to everlasting you are God.

Psalms 95, 4 verses

3 For the LORD is the great God, the great King above all gods.

4 In his hand are the depths of the earth, and the mountain peaks belong to him.

5 The sea is his, for he made it, and his hands formed the dry land.

6 Come, let us bow down in worship, let us kneel before the LORD our Maker;

Psalms 96, 2 verses

4 For great is the LORD and most worthy of praise; he is to be feared above all gods.

5 For all the gods of the nations are idols, but the LORD made the heavens.

Psalms 100, 1 verse

3 Know that the LORD is God. It is he who made us, and we are his; we are his people, the sheep of his pasture.

Psalms 103, 2 verses

13 As a father has compassion on his children, so the LORD has compassion on those who fear him;

14 for he knows how we are formed, he remembers that we are dust.

Psalms 104, 9 verses

1 Praise the LORD, O my soul. O LORD my God, you are very great; you are clothed with splendor and majesty.

2 He wraps himself in light as with a garment; he stretches out the heavens like a tent

3 and lays the beams of his upper chambers on their waters. He makes the clouds his chariot and rides on the wings of the wind.

4 He makes winds his messengers, flames of fire his servants.

5 He set the earth on its foundations; it can never be moved.

6 You covered it with the deep as with a garment; the waters stood above the mountains.

7 But at your rebuke the waters fled, at the sound of your thunder they took to flight;

8 they flowed over the mountains, they went down into the valleys, to the place you assigned for them.

9 You set a boundary they cannot cross; never again will they cover the earth.

Psalms 115, 1 verse

16 The highest heavens belong to the LORD, but the earth he has given to man.

Psalms 118, 1 verse

24 This is the day the LORD has made; let us rejoice and be glad in it.

Psalms 119, 1 verse

90 Your faithfulness continues through all generations; you established the earth, and it endures.

Psalms 121, 1 verse

2 My help comes from the LORD, the Maker of heaven and earth.

Psalms 135, 1 verse

7 He makes clouds rise from the ends of the earth; he sends lightning with the rain and brings out the wind from his storehouses.

Psalms 136, 7 verses

5 who by his understanding made the heavens, His love endures forever.

6 who spread out the earth upon the waters, His love endures forever.

7 who made the great lights-- His love endures forever.

8 the sun to govern the day, His love endures forever.

9 the moon and stars to govern the night; His love endures forever.

25 and who gives food to every creature. His love endures forever.

26 Give thanks to the God of heaven. His love endures forever.

Psalms 138, 1 verse

8 The LORD will fulfill [his purpose] for me; your love, O LORD, endures forever-- do not abandon the works of your hands.

Psalms 139, 4 verses

13 For you created my inmost being; you knit me together in my mother's womb.

14 I praise you because I am fearfully and wonderfully made; your works are wonderful, I know that full well.

15 My frame was not hidden from you when I was made in the secret place. When I was woven together in the depths of the earth,

16 your eyes saw my unformed body. All the days ordained for me were written in your book before one of them came to be.

Psalms 145, 2 verses

15 The eyes of all look to you, and you give them their food at the proper time.

16 You open your hand and satisfy the desires of every living thing.

Psalms 146, 2 verses

5 Blessed is he whose help is the God of Jacob, whose hope is in the LORD his God,

6 the Maker of heaven and earth, the sea, and everything in them-- the LORD, who remains faithful forever.

Psalms 147, 7 verses

4 He determines the number of the stars and calls them each by name.

8 He covers the sky with clouds; he supplies the earth with rain and makes grass grow on the hills.

9 He provides food for the cattle and for the young ravens when they call.

15 He sends his command to the earth; his word runs swiftly.

16 He spreads the snow like wool and scatters the frost like ashes.

17 He hurls down his hail like pebbles. Who can withstand his icy blast?

18 He sends his word and melts them; he stirs up his breezes, and the waters flow.

Proverbs Total: 11 Verses

Proverbs 8, 10 verses

22 "The LORD brought me forth as the first of his works, before his deeds of old;

23 I was appointed from eternity, from the beginning, before the world began.

24 When there were no oceans, I was given birth, when there were no springs abounding with water;

25 before the mountains were settled in place, before the hills, I was given birth,

26 before he made the earth or its fields or any of the dust of the world.

27 I was there when he set the heavens in place, when he marked out the horizon on the face of the deep,

28 when he established the clouds above and fixed securely the fountains of the deep,

29 when he gave the sea its boundary so the waters would not overstep his command, and when he marked out the foundations of the earth.

30 Then I was the craftsman at his side. I was filled with delight day after day, rejoicing always in his presence,

31 rejoicing in his whole world and delighting in mankind.

Proverbs 29, 1 verse

13 The poor man and the oppressor have this in common: The LORD gives sight to the eyes of both.

Ecclesiastes Total: 3 Verses

Ecclesiastes 3, 2 verses

11 He has made everything beautiful in its time. He has also set eternity in the hearts of men; yet they cannot fathom what God has done from beginning to end.

14 I know that everything God does will endure forever; nothing can be added to it and nothing taken from it. God does it so that men will revere him.

Ecclesiastes 11, 1 verse

5 As you do not know the path of the wind, or how the body is formed in a mother's womb, so you cannot understand the work of God, the Maker of all things.

Isaiah Total: 33 Verses

Isaiah 22, 1 verse

11 You built a reservoir between the two walls for the water of the Old Pool, but you did not look to the One who made it, or have regard for the One who planned it long ago.

Isaiah 34, 7 verses

11 The desert owl and screech owl will possess it; the great owl and the raven will nest there. God will stretch out over Edom the measuring line of chaos and the plumb line of desolation.

12 Her nobles will have nothing there to be called a kingdom, all her princes will vanish away.

13 Thorns will overrun her citadels, nettles and brambles her strongholds. She will become a haunt for jackals, a home for owls.

14 Desert creatures will meet with hyenas, and wild goats will bleat to each other; there the night creatures will also repose and find for themselves places of rest.

15 The owl will nest there and lay eggs, she will hatch them, and care for her young under the shadow of her wings; there also the falcons will gather, each with its mate.

16 Look in the scroll of the LORD and read: None of these will be missing, not one will lack her mate. For it is his mouth that has given the order, and his Spirit will gather them together.

17 He allots their portions; his hand distributes them by measure. They will possess it forever and dwell there from generation to generation.

Isaiah 40, 2 verses

22 He sits enthroned above the circle of the earth, and its people are like grasshoppers. He stretches out the heavens like a canopy, and spreads them out like a tent to live in.

26 Lift your eyes and look to the heavens: Who created all these? He who brings out the starry host one by one, and calls them each by name. Because of his great power and mighty strength, not one of them is missing.

Isaiah 42, 1 verse

5 This is what God the LORD says-- he who created the heavens and stretched them out, who spread out the earth and all that comes out of it, who gives breath to its people, and life to those who walk on it:

Isaiah 43, 1 verse

1 But now, this is what the LORD says-- he who created you, O Jacob, he who formed you, O Israel: "Fear not, for I have redeemed you; I have summoned you by name; you are mine.

Isaiah 44, 1 verse

2 This is what the LORD says-- he who made you, who formed you in the womb, and who will help you: Do not be afraid, O Jacob, my servant, Jeshurun, whom I have chosen.

Isaiah 45, 8 verses

5 I am the LORD, and there is no other; apart from me there is no God. I will strengthen you, though you have not acknowledged me,

6 so that from the rising of the sun to the place of its setting men may know there is none besides me. I am the LORD, and there is no other.

7 I form the light and create darkness, I bring prosperity and create disaster; I, the LORD, do all these things.

8 "You heavens above, rain down righteousness; let the clouds shower it down. Let the earth open wide, let salvation spring up, let righteousness grow with it; I, the LORD, have created it.

9 "Woe to him who quarrels with his Maker, to him who is but a potsherd among the potsherds on the ground. Does the clay say to the potter, 'What are you making?' Does your work say, 'He has no hands'?

10 Woe to him who says to his father, 'What have you begotten?' or to his mother, 'What have you brought to birth?'

11 "This is what the LORD says-- the Holy One of Israel, and its Maker: Concerning things to come, do you question me about my children, or give me orders about the work of my hands?

12 It is I who made the earth and created mankind upon it. My own hands stretched out the heavens; I marshaled their starry hosts.

Isaiah 46, 1 verse

11 From the east I summon a bird of prey; from a far-off land, a man to fulfill my purpose. What I have said, that will I bring about; what I have planned, that will I do.

Isaiah 48, 1 verse

13 My own hand laid the foundations of the earth, and my right hand spread out the heavens; when I summon them, they all stand up together.

Isaiah 49, 2 verses

1 Listen to me, you islands; hear this, you distant nations: Before I was born the LORD called me; from my birth he has made mention of my name.

2 He made my mouth like a sharpened sword, in the shadow of his hand he hid me; he made me into a polished arrow and concealed me in his quiver.

Isaiah 50, 1 verse

3 I clothe the sky with darkness and make sackcloth its covering."

Isaiah 51, 2 verses

15 For I am the LORD your God, who churns up the sea so that its waves roar-- the LORD Almighty is his name.

16 I have put my words in your mouth and covered you with the shadow of my hand-- I who set the heavens in place, who laid the foundations of the earth, and who say to Zion, 'You are my people.'"

Isaiah 54, 1 verse
16 "See, it is I who created the blacksmith who fans the coals into flame and forges a weapon fit for its work. And it is I who have created the destroyer to work havoc;

Isaiah 57, 1 verse
16 I will not accuse forever, nor will I always be angry, for then the spirit of man would grow faint before me-- the breath of man that I have created.

Isaiah 64, 1 verse
8 Yet, O LORD, you are our Father. We are the clay, you are the potter; we are all the work of your hand.

Isaiah 66, 2 verses
1 This is what the LORD says: "Heaven is my throne, and the earth is my footstool. Where is the house you will build for me? Where will my resting place be?
2 Has not my hand made all these things, and so they came into being?" declares the LORD. "This is the one I esteem: he who is humble and contrite in spirit, and trembles at my word.

Jeremiah Total: 12 Verses

Jeremiah 1, 1 verse
5 "Before I formed you in the womb I knew you, before you were born I set you apart; I appointed you as a prophet to the nations."

Jeremiah 3, 1 verse
3 Therefore the showers have been withheld, and no spring rains have fallen. Yet you have the brazen look of a prostitute; you refuse to blush with shame.

Jeremiah 5, 1 verse
22 Should you not fear me?" declares the LORD. "Should you not tremble in my presence? I made the sand a boundary for the sea, an everlasting barrier it cannot cross. The waves may roll, but they cannot prevail; they may roar, but they cannot cross it.

Jeremiah 10, 2 verses
12 But God made the earth by his power; he founded the world by his wisdom and stretched out the heavens by his understanding.

13 When he thunders, the waters in the heavens roar; he makes clouds rise from the ends of the earth. He sends lightning with the rain and brings out the wind from his storehouses.

Jeremiah 14, 1 verse
22 Do any of the worthless idols of the nations bring rain? Do the skies themselves send down showers? No, it is you, O LORD our God. Therefore our hope is in you, for you are the one who does all this.

Jeremiah 27, 1 verse
5 With my great power and outstretched arm I made the earth and its people and the animals that are on it, and I give it to anyone I please.

Jeremiah 31, 3 verses
35 This is what the LORD says, he who appoints the sun to shine by day, who decrees the moon and stars to shine by night, who stirs up the sea so that its waves roar-- the LORD Almighty is his name:
36 "Only if these decrees vanish from my sight," declares the LORD, "will the descendants of Israel ever cease to be a nation before me."
37 This is what the LORD says: "Only if the heavens above can be measured and the foundations of the earth below be searched out will I reject all the descendants of Israel because of all they have done," declares the LORD.

Jeremiah 51, 2 verses
15 "He made the earth by his power; he founded the world by his wisdom and stretched out the heavens by his understanding.
16 When he thunders, the waters in the heavens roar; he makes clouds rise from the ends of the earth. He sends lightning with the rain and brings out the wind from his storehouses.

Mathew Total: 17 Verses

Mathew 2, 3 verses
13 When they had gone, an angel of the Lord appeared to Joseph in a dream. "Get up," he said, "take the child and his mother and escape to Egypt. Stay there until I tell you, for Herod is going to search for the child to kill him."
14 So he got up, took the child and his mother during the night and left for Egypt,
15 where he stayed until the death of Herod. And so was fulfilled what the Lord had said through the prophet: "Out of Egypt I called my son."

Mathew 3, 2 verses

16 As soon as Jesus was baptized, he went up out of the water. At that moment heaven was opened, and he saw the Spirit of God descending like a dove and lighting on him.

17 And a voice from heaven said, "This is my Son, whom I love; with him I am well pleased."

Mathew 5, 1 verse

45 that you may be sons of your Father in heaven. He causes his sun to rise on the evil and the good, and sends rain on the righteous and the unrighteous.

Mathew 6, 4 verses

6 But when you pray, go into your room, close the door and pray to your Father, who is unseen. Then your Father, who sees what is done in secret, will reward you.

18 so that it will not be obvious to men that you are fasting, but only to your Father, who is unseen; and your Father, who sees what is done in secret, will reward you.

26 Look at the birds of the air; they do not sow or reap or store away in barns, and yet your heavenly Father feeds them. Are you not much more valuable than they?

30 If that is how God clothes the grass of the field, which is here today and tomorrow is thrown into the fire, will he not much more clothe you, O you of little faith?

Mathew 7, 3 verses

7 "Ask and it will be given to you; seek and you will find; knock and the door will be opened to you.

8 For everyone who asks receives; he who seeks finds; and to him who knocks, the door will be opened.

11 If you, then, though you are evil, know how to give good gifts to your children, how much more will your Father in heaven give good gifts to those who ask him!

Mathew 10, 2 verses

29 Are not two sparrows sold for a penny? Yet not one of them will fall to the ground apart from the will of your Father.

30 And even the very hairs of your head are all numbered.

Mathew 15, 1 verse

13 He replied, "Every plant that my heavenly Father has not planted will be pulled up by the roots.

Mathew 16, 1 verse

17 Jesus replied, "Blessed are you, Simon son of Jonah, for this was not revealed to you by man, but by my Father in heaven.

Mark Total: 2 Verses

Mark 1, 1 verse
11 And a voice came from heaven: "You are my Son, whom I love; with you I am well pleased."

Mark 5, 1 verse
19 Jesus did not let him, but said, "Go home to your family and tell them how much the Lord has done for you, and how he has had mercy on you."

Luke Total: 50 Verses

Luke 1, 32 verses
19 The angel answered, "I am Gabriel. I stand in the presence of God, and I have been sent to speak to you and to tell you this good news.
25 "The Lord has done this for me," she said. "In these days he has shown his favor and taken away my disgrace among the people."
26 In the sixth month, God sent the angel Gabriel to Nazareth, a town in Galilee,
27 to a virgin pledged to be married to a man named Joseph, a descendant of David. The virgin's name was Mary.
28 The angel went to her and said, "Greetings, you who are highly favored! The Lord is with you."
29 Mary was greatly troubled at his words and wondered what kind of greeting this might be.
30 But the angel said to her, "Do not be afraid, Mary, you have found favor with God.
31 You will be with child and give birth to a son, and you are to give him the name Jesus.
32 He will be great and will be called the Son of the Most High. The Lord God will give him the throne of his father David,
33 and he will reign over the house of Jacob forever; his kingdom will never end."
36 Even Elizabeth your relative is going to have a child in her old age, and she who was said to be barren is in her sixth month.
37 For nothing is impossible with God."
38 "I am the Lord's servant," Mary answered. "May it be to me as you have said." Then the angel left her.
46 And Mary said: "My soul glorifies the Lord

47 and my spirit rejoices in God my Savior,

48 for he has been mindful of the humble state of his servant. From now on all generations will call me blessed,

49 for the Mighty One has done great things for me-- holy is his name.

50 His mercy extends to those who fear him, from generation to generation.

51 He has performed mighty deeds with his arm; he has scattered those who are proud in their inmost thoughts.

52 He has brought down rulers from their thrones but has lifted up the humble.

53 He has filled the hungry with good things but has sent the rich away empty.

54 He has helped his servant Israel, remembering to be merciful

55 to Abraham and his descendants forever, even as he said to our fathers."

67 His father Zechariah was filled with the Holy Spirit and prophesied:

68 "Praise be to the Lord, the God of Israel, because he has come and has redeemed his people.

69 He has raised up a horn of salvation for us in the house of his servant David

70 (as he said through his holy prophets of long ago),

71 salvation from our enemies and from the hand of all who hate us--

72 to show mercy to our fathers and to remember his holy covenant,

73 the oath he swore to our father Abraham:

74 to rescue us from the hand of our enemies, and to enable us to serve him without fear

75 in holiness and righteousness before him all our days.

Luke 2, 5 verses

28 Simeon took him in his arms and praised God, saying:

29 "Sovereign Lord, as you have promised, you now dismiss your servant in peace.

30 For my eyes have seen your salvation,

31 which you have prepared in the sight of all people,

32 a light for revelation to the Gentiles and for glory to your people Israel."

Luke 3, 2 verses

21 When all the people were being baptized, Jesus was baptized too. And as he was praying, heaven was opened

22 and the Holy Spirit descended on him in bodily form like a dove. And a voice came from heaven: "You are my Son, whom I love; with you I am well pleased."

Luke 9, 1 verse

35 A voice came from the cloud, saying, "This is my Son, whom I have chosen; listen to him."

Luke 11, 2 verses

13 If you then, though you are evil, know how to give good gifts to your children, how much more will your Father in heaven give the Holy Spirit to those who ask him!"

49 Because of this, God in his wisdom said, 'I will send them prophets and apostles, some of whom they will kill and others they will persecute.'

Luke 12, 8 verses

20 "But God said to him, 'You fool! This very night your life will be demanded from you. Then who will get what you have prepared for yourself?'

24 Consider the ravens: They do not sow or reap, they have no storeroom or barn; yet God feeds them. And how much more valuable you are than birds!

27 "Consider how the lilies grow. They do not labor or spin. Yet I tell you, not even Solomon in all his splendor was dressed like one of these.

28 If that is how God clothes the grass of the field, which is here today, and tomorrow is thrown into the fire, how much more will he clothe you, O you of little faith!

29 And do not set your heart on what you will eat or drink; do not worry about it.

30 For the pagan world runs after all such things, and your Father knows that you need them.

31 But seek his kingdom, and these things will be given to you as well.

32 "Do not be afraid, little flock, for your Father has been pleased to give you the kingdom.

John Total: 57 Verses

John 1, 1 verse

6 There came a man who was sent from God; his name was John.

John 5, 12 verses

17 Jesus said to them, "My Father is always at his work to this very day, and I, too, am working."

19 Jesus gave them this answer: "I tell you the truth, the Son can do nothing by himself; he can do only what he sees his Father doing, because whatever the Father does the Son also does.

20 For the Father loves the Son and shows him all he does. Yes, to your amazement he will show him even greater things than these.

21 For just as the Father raises the dead and gives them life, even so the Son gives life to whom he is pleased to give it.

22 Moreover, the Father judges no one, but has entrusted all judgment to the Son,

23 that all may honor the Son just as they honor the Father. He who does not honor the Son does not honor the Father, who sent him.

26 For as the Father has life in himself, so he has granted the Son to have life in himself.

27 And he has given him authority to judge because he is the Son of Man.

30 By myself I can do nothing; I judge only as I hear, and my judgment is just, for I seek not to please myself but him who sent me.

36 "I have testimony weightier than that of John. For the very work that the Father has given me to finish, and which I am doing, testifies that the Father has sent me.

37 And the Father who sent me has himself testified concerning me. You have never heard his voice nor seen his form,

38 nor does his word dwell in you, for you do not believe the one he sent.

John 6, 16 verses

27 Do not work for food that spoils, but for food that endures to eternal life, which the Son of Man will give you. On him God the Father has placed his seal of approval."

28 Then they asked him, "What must we do to do the works God requires?"

29 Jesus answered, "The work of God is this: to believe in the one he has sent."

30 So they asked him, "What miraculous sign then will you give that we may see it and believe you? What will you do?

31 Our forefathers ate the manna in the desert; as it is written: 'He gave them bread from heaven to eat.'"

32 Jesus said to them, "I tell you the truth, it is not Moses who has given you the bread from heaven, but it is my Father who gives you the true bread from heaven.

33 For the bread of God is he who comes down from heaven and gives life to the world."

34 "Sir," they said, "from now on give us this bread."

35 Then Jesus declared, "I am the bread of life. He who comes to me will never go hungry, and he who believes in me will never be thirsty.

37 All that the Father gives me will come to me, and whoever comes to me I will never drive away.

38 For I have come down from heaven not to do my will but to do the will of him who sent me.

39 And this is the will of him who sent me, that I shall lose none of all that he has given me, but raise them up at the last day.

40 For my Father's will is that everyone who looks to the Son and believes in him shall have eternal life, and I will raise him up at the last day."

44 "No one can come to me unless the Father who sent me draws him, and I will raise him up at the last day.

45 It is written in the Prophets: 'They will all be taught by God.' Everyone who listens to the Father and learns from him comes to me.

65 He went on to say, "This is why I told you that no one can come to me unless the Father has enabled him."

John 11, 2 verses

41 So they took away the stone. Then Jesus looked up and said, "Father, I thank you that you have heard me.

42 I knew that you always hear me, but I said this for the benefit of the people standing here, that they may believe that you sent me."

John 12, 4 verses

28 Father, glorify your name!" Then a voice came from heaven, "I have glorified it, and will glorify it again."

45 When he looks at me, he sees the one who sent me.

49 For I did not speak of my own accord, but the Father who sent me commanded me what to say and how to say it.

50 I know that his command leads to eternal life. So whatever I say is just what the Father has told me to say."

John 14, 12 verses

6 Jesus answered, "I am the way and the truth and the life. No one comes to the Father except through me.

7 If you really knew me, you would know my Father as well. From now on, you do know him and have seen him."

8 Philip said, "Lord, show us the Father and that will be enough for us."

9 Jesus answered: "Don't you know me, Philip, even after I have been among you such a long time? Anyone who has seen me has seen the Father. How can you say, 'Show us the Father'?

10 Don't you believe that I am in the Father, and that the Father is in me? The words I say to you are not just my own. Rather, it is the Father, living in me, who is doing his work.

11 Believe me when I say that I am in the Father and the Father is in me; or at least believe on the evidence of the miracles themselves.

12 I tell you the truth, anyone who has faith in me will do what I have been doing. He will do even greater things than these, because I am going to the Father.

13 And I will do whatever you ask in my name, so that the Son may bring glory to the Father.

14 You may ask me for anything in my name, and I will do it.

16 And I will ask the Father, and he will give you another Counselor to be with you forever--

17 the Spirit of truth. The world cannot accept him, because it neither sees him nor knows him. But you know him, for he lives with you and will be in you.

26 But the Counselor, the Holy Spirit, whom the Father will send in my name, will teach you all things and will remind you of everything I have said to you.

John 16, 10 verses

23 In that day you will no longer ask me anything. I tell you the truth, my Father will give you whatever you ask in my name.

24 Until now you have not asked for anything in my name. Ask and you will receive, and your joy will be complete.

25 "Though I have been speaking figuratively, a time is coming when I will no longer use this kind of language but will tell you plainly about my Father.

26 In that day you will ask in my name. I am not saying that I will ask the Father on your behalf.

27 No, the Father himself loves you because you have loved me and have believed that I came from God.

28 I came from the Father and entered the world; now I am leaving the world and going back to the Father."

29 Then Jesus' disciples said, "Now you are speaking clearly and without figures of speech.

30 Now we can see that you know all things and that you do not even need to have anyone ask you questions. This makes us believe that you came from God."

31 "You believe at last!" Jesus answered.

32 "But a time is coming, and has come, when you will be scattered, each to his own home. You will leave me all alone. Yet I am not alone, for my Father is with me.

Acts Total: 89 Verses

Acts 2, 6 verses

22 "Men of Israel, listen to this: Jesus of Nazareth was a man accredited by God to you by miracles, wonders and signs, which God did among you through him, as you yourselves know.

23 This man was handed over to you by God's set purpose and foreknowledge; and you, with the help of wicked men, put him to death by nailing him to the cross.

24 But God raised him from the dead, freeing him from the agony of death, because it was impossible for death to keep its hold on him.

30 But he was a prophet and knew that God had promised him on oath that he would place one of his descendants on his throne.

31 Seeing what was ahead, he spoke of the resurrection of the Christ, that he was not abandoned to the grave, nor did his body see decay.

32 God has raised this Jesus to life, and we are all witnesses of the fact.

Acts 3, 14 verses

13 The God of Abraham, Isaac and Jacob, the God of our fathers, has glorified his servant Jesus. You handed him over to be killed, and you disowned him before Pilate, though he had decided to let him go.

14 You disowned the Holy and Righteous One and asked that a murderer be released to you.

15 You killed the author of life, but God raised him from the dead. We are witnesses of this.

16 By faith in the name of Jesus, this man whom you see and know was made strong. It is Jesus' name and the faith that comes through him that has given this complete healing to him, as you can all see.

17 "Now, brothers, I know that you acted in ignorance, as did your leaders.

18 But this is how God fulfilled what he had foretold through all the prophets, saying that his Christ would suffer.

19 Repent, then, and turn to God, so that your sins may be wiped out, that times of refreshing may come from the Lord,

20 and that he may send the Christ, who has been appointed for you-- even Jesus.

21 He must remain in heaven until the time comes for God to restore everything, as he promised long ago through his holy prophets.

22 For Moses said, 'The Lord your God will raise up for you a prophet like me from among your own people; you must listen to everything he tells you.

23 Anyone who does not listen to him will be completely cut off from among his people.'

24 "Indeed, all the prophets from Samuel on, as many as have spoken, have foretold these days.

25 And you are heirs of the prophets and of the covenant God made with your fathers. He said to Abraham, 'Through your offspring all peoples on earth will be blessed.'

26 When God raised up his servant, he sent him first to you to bless you by turning each of you from your wicked ways."

Acts 4, 1 verse

10 then know this, you and all the people of Israel: It is by the name of Jesus Christ of Nazareth, whom you crucified but whom God raised from the dead, that this man stands before you healed.

Acts 5, 2 verses

30 The God of our fathers raised Jesus from the dead-- whom you had killed by hanging him on a tree.

31 God exalted him to his own right hand as Prince and Savior that he might give repentance and forgiveness of sins to Israel.

Acts 7, 2 verses

49 "'Heaven is my throne, and the earth is my footstool. What kind of house will you build for me? says the Lord. Or where will my resting place be?

50 Has not my hand made all these things?'

Acts 10, 13 verses

28 He said to them: "You are well aware that it is against our law for a Jew to associate with a Gentile or visit him. But God has shown me that I should not call any man impure or unclean.

30 Cornelius answered: "Four days ago I was in my house praying at this hour, at three in the afternoon. Suddenly a man in shining clothes stood before me

31 and said, 'Cornelius, God has heard your prayer and remembered your gifts to the poor.

34 Then Peter began to speak: "I now realize how true it is that God does not show favoritism

35 but accepts men from every nation who fear him and do what is right.

36 You know the message God sent to the people of Israel, telling the good news of peace through Jesus Christ, who is Lord of all.

37 You know what has happened throughout Judea, beginning in Galilee after the baptism that John preached--

38 how God anointed Jesus of Nazareth with the Holy Spirit and power, and how he went around doing good and healing all who were under the power of the devil, because God was with him.

39 "We are witnesses of everything he did in the country of the Jews and in Jerusalem. They killed him by hanging him on a tree,

40 but God raised him from the dead on the third day and caused him to be seen.

41 He was not seen by all the people, but by witnesses whom God had already chosen-- by us who ate and drank with him after he rose from the dead.

42 He commanded us to preach to the people and to testify that he is the one whom God appointed as judge of the living and the dead.

43 All the prophets testify about him that everyone who believes in him receives forgiveness of sins through his name."

Acts 11, 3 verses

17 So if God gave them the same gift as he gave us, who believed in the Lord Jesus Christ, who was I to think that I could oppose God?"

18 When they heard this, they had no further objections and praised God, saying, "So then, God has granted even the Gentiles repentance unto life."

21 The Lord's hand was with them, and a great number of people believed and turned to the Lord.

Acts 12, 7 verses

7 Suddenly an angel of the Lord appeared and a light shone in the cell. He struck Peter on the side and woke him up. "Quick, get up!" he said, and the chains fell off Peter's wrists.

8 Then the angel said to him, "Put on your clothes and sandals." And Peter did so. "Wrap your cloak around you and follow me," the angel told him.

9 Peter followed him out of the prison, but he had no idea that what the angel was doing was really happening; he thought he was seeing a vision.

10 They passed the first and second guards and came to the iron gate leading to the city. It opened for them by itself, and they went through it. When they had walked the length of one street, suddenly the angel left him.

11 Then Peter came to himself and said, "Now I know without a doubt that the Lord sent his angel and rescued me from Herod's clutches and from everything the Jewish people were anticipating."

22 They shouted, "This is the voice of a god, not of a man."

23 Immediately, because Herod did not give praise to God, an angel of the Lord struck him down, and he was eaten by worms and died.

Acts 13, 17 verses

2 While they were worshiping the Lord and fasting, the Holy Spirit said, "Set apart for me Barnabas and Saul for the work to which I have called them."

17 The God of the people of Israel chose our fathers; he made the people prosper during their stay in Egypt, with mighty power he led them out of that country,

18 he endured their conduct for about forty years in the desert,

19 he overthrew seven nations in Canaan and gave their land to his people as their inheritance.

20 All this took about 450 years. "After this, God gave them judges until the time of Samuel the prophet.

21 Then the people asked for a king, and he gave them Saul son of Kish, of the tribe of Benjamin, who ruled forty years.

22 After removing Saul, he made David their king. He testified concerning him: 'I have found David son of Jesse a man after my own heart; he will do everything I want him to do.'

23 "From this man's descendants God has brought to Israel the Savior Jesus, as he promised.

29 When they had carried out all that was written about him, they took him down from the tree and laid him in a tomb.

30 But God raised him from the dead,

31 and for many days he was seen by those who had traveled with him from Galilee to Jerusalem. They are now his witnesses to our people.

32 "We tell you the good news: What God promised our fathers

33 he has fulfilled for us, their children, by raising up Jesus. As it is written in the second Psalm: "'You are my Son; today I have become your Father.'

34 The fact that God raised him from the dead, never to decay, is stated in these words: "'I will give you the holy and sure blessings promised to David.'

35 So it is stated elsewhere: "'You will not let your Holy One see decay.'

36 "For when David had served God's purpose in his own generation, he fell asleep; he was buried with his fathers and his body decayed.

37 But the one whom God raised from the dead did not see decay.

Acts 14, 4 verses

15 "Men, why are you doing this? We too are only men, human like you. We are bringing you good news, telling you to turn from these worthless things to the living God, who made heaven and earth and sea and everything in them.

16 In the past, he let all nations go their own way.

17 Yet he has not left himself without testimony: He has shown kindness by giving you rain from heaven and crops in their seasons; he provides you with plenty of food and fills your hearts with joy."

27 On arriving there, they gathered the church together and reported all that God had done through them and how he had opened the door of faith to the Gentiles.

Acts 15, 12 verses

7 After much discussion, Peter got up and addressed them: "Brothers, you know that some time ago God made a choice among you that the Gentiles might hear from my lips the message of the gospel and believe.

8 God, who knows the heart, showed that he accepted them by giving the Holy Spirit to them, just as he did to us.

9 He made no distinction between us and them, for he purified their hearts by faith.

10 Now then, why do you try to test God by putting on the necks of the disciples a yoke that neither we nor our fathers have been able to bear?

11 No! We believe it is through the grace of our Lord Jesus that we are saved, just as they are."

12 The whole assembly became silent as they listened to Barnabas and Paul telling about the miraculous signs and wonders God had done among the Gentiles through them.

13 When they finished, James spoke up: "Brothers, listen to me.

14 Simon has described to us how God at first showed his concern by taking from the Gentiles a people for himself.

15 The words of the prophets are in agreement with this, as it is written:

16 "'After this I will return and rebuild David's fallen tent. Its ruins I will rebuild, and I will restore it,

17 that the remnant of men may seek the Lord, and all the Gentiles who bear my name, says the Lord, who does these things'

18 that have been known for ages.

Acts 16, 1 verse

14 One of those listening was a woman named Lydia, a dealer in purple cloth from the city of Thyatira, who was a worshiper of God. The Lord opened her heart to respond to Paul's message.

Acts 19, 2 verses

11 God did extraordinary miracles through Paul,

12 so that even handkerchiefs and aprons that had touched him were taken to the sick, and their illnesses were cured and the evil spirits left them.

Acts 22, 1 verse

14 "Then he said: 'The God of our fathers has chosen you to know his will and to see the Righteous One and to hear words from his mouth.

Acts 23, 1 verse

11 The following night the Lord stood near Paul and said, "Take courage! As you have testified about me in Jerusalem, so you must also testify in Rome."

Acts 27, 3 verses

23 Last night an angel of the God whose I am and whom I serve stood beside me

24 and said, 'Do not be afraid, Paul. You must stand trial before Caesar; and God has graciously given you the lives of all who sail with you.'

25 So keep up your courage, men, for I have faith in God that it will happen just as he told me.

Romans Total: 80 Verses

Romans 1, 12 verses

17 For in the gospel a righteousness from God is revealed, a righteousness that is by faith from first to last, just as it is written: "The righteous will live by faith."

18 The wrath of God is being revealed from heaven against all the godlessness and wickedness of men who suppress the truth by their wickedness,

19 since what may be known about God is plain to them, because God has made it plain to them.

20 For since the creation of the world God's invisible qualities-- his eternal power and divine nature-- have been clearly seen, being understood from what has been made, so that men are without excuse.

21 For although they knew God, they neither glorified him as God nor gave thanks to him, but their thinking became futile and their foolish hearts were darkened.

22 Although they claimed to be wise, they became fools

23 and exchanged the glory of the immortal God for images made to look like mortal man and birds and animals and reptiles.

24 Therefore God gave them over in the sinful desires of their hearts to sexual impurity for the degrading of their bodies with one another.

25 They exchanged the truth of God for a lie, and worshiped and served created things rather than the Creator-- who is forever praised. Amen.

26 Because of this, God gave them over to shameful lusts. Even their women exchanged natural relations for unnatural ones.

27 In the same way the men also abandoned natural relations with women and were inflamed with lust for one another. Men committed indecent acts with other men, and received in themselves the due penalty for their perversion.

28 Furthermore, since they did not think it worthwhile to retain the knowledge of God, he gave them over to a depraved mind, to do what ought not to be done.

Romans 2, 15 verses

2 Now we know that God's judgment against those who do such things is based on truth.

3 So when you, a mere man, pass judgment on them and yet do the same things, do you think you will escape God's judgment?

4 Or do you show contempt for the riches of his kindness, tolerance and patience, not realizing that God's kindness leads you toward repentance?

5 But because of your stubbornness and your unrepentant heart, you are storing up wrath against yourself for the day of God's wrath, when his righteous judgment will be revealed.

6 God "will give to each person according to what he has done."

7 To those who by persistence in doing good seek glory, honor and immortality, he will give eternal life.

8 But for those who are self-seeking and who reject the truth and follow evil, there will be wrath and anger.

9 There will be trouble and distress for every human being who does evil: first for the Jew, then for the Gentile;

10 but glory, honor and peace for everyone who does good: first for the Jew, then for the Gentile.

11 For God does not show favoritism.

12 All who sin apart from the law will also perish apart from the law, and all who sin under the law will be judged by the law.

13 For it is not those who hear the law who are righteous in God's sight, but it is those who obey the law who will be declared righteous.

14 (Indeed, when Gentiles, who do not have the law, do by nature things required by the law, they are a law for themselves, even though they do not have the law,

15 since they show that the requirements of the law are written on their hearts, their consciences also bearing witness, and their thoughts now accusing, now even defending them.)

16 This will take place on the day when God will judge men's secrets through Jesus Christ, as my gospel declares.

Romans 3, 3 verses

1 What advantage, then, is there in being a Jew, or what value is there in circumcision?

2 Much in every way! First of all, they have been entrusted with the very words of God.

25 God presented him as a sacrifice of atonement, through faith in his blood. He did this to demonstrate his justice, because in his forbearance he had left the sins committed beforehand unpunished--

Romans 4, 4 verses

22 This is why "it was credited to him as righteousness."

23 The words "it was credited to him" were written not for him alone,

24 but also for us, to whom God will credit righteousness-- for us who believe in him who raised Jesus our Lord from the dead.

25 He was delivered over to death for our sins and was raised to life for our justification.

Romans 5, 4 verses

8 But God demonstrates his own love for us in this: While we were still sinners, Christ died for us.

9 Since we have now been justified by his blood, how much more shall we be saved from God's wrath through him!

10 For if, when we were God's enemies, we were reconciled to him through the death of his Son, how much more, having been reconciled, shall we be saved through his life!

11 Not only is this so, but we also rejoice in God through our Lord Jesus Christ, through whom we have now received reconciliation.

Romans 8, 10 verses

19 The creation waits in eager expectation for the sons of God to be revealed.

20 For the creation was subjected to frustration, not by its own choice, but by the will of the one who subjected it, in hope

21 that the creation itself will be liberated from its bondage to decay and brought into the glorious freedom of the children of God.

22 We know that the whole creation has been groaning as in the pains of childbirth right up to the present time.

23 Not only so, but we ourselves, who have the first fruits of the Spirit, groan inwardly as we wait eagerly for our adoption as sons, the redemption of our bodies.

35 Who shall separate us from the love of Christ? Shall trouble or hardship or persecution or famine or nakedness or danger or sword?

36 As it is written: "For your sake we face death all day long; we are considered as sheep to be slaughtered."

37 No, in all these things we are more than conquerors through him who loved us.

38 For I am convinced that neither death nor life, neither angels nor demons, neither the present nor the future, nor any powers,

39 neither height nor depth, nor anything else in all creation, will be able to separate us from the love of God that is in Christ Jesus our Lord.

Romans 9, 5 verses

17 For the Scripture says to Pharaoh: "I raised you up for this very purpose, that I might display my power in you and that my name might be proclaimed in all the earth."

18 Therefore God has mercy on whom he wants to have mercy, and he hardens whom he wants to harden.

19 One of you will say to me: "Then why does God still blame us? For who resists his will?"

20 But who are you, O man, to talk back to God? "Shall what is formed say to him who formed it, 'Why did you make me like this?'"

21 Does not the potter have the right to make out of the same lump of clay some pottery for noble purposes and some for common use?

Romans 11, 19 verses

2 God did not reject his people, whom he foreknew. Don't you know what the Scripture says in the passage about Elijah-- how he appealed to God against Israel:

3 "Lord, they have killed your prophets and torn down your altars; I am the only one left, and they are trying to kill me"?

4 And what was God's answer to him? "I have reserved for myself seven thousand who have not bowed the knee to Baal."

5 So too, at the present time there is a remnant chosen by grace.

6 And if by grace, then it is no longer by works; if it were, grace would no longer be grace.

7 What then? What Israel sought so earnestly it did not obtain, but the elect did. The others were hardened,

8 as it is written: "God gave them a spirit of stupor, eyes so that they could not see and ears so that they could not hear, to this very day."

13 I am talking to you Gentiles. Inasmuch as I am the apostle to the Gentiles, I make much of my ministry

14 in the hope that I may somehow arouse my own people to envy and save some of them.

15 For if their rejection is the reconciliation of the world, what will their acceptance be but life from the dead?

16 If the part of the dough offered as first fruits is holy, then the whole batch is holy; if the root is holy, so are the branches.

17 If some of the branches have been broken off, and you, though a wild olive shoot, have been grafted in among the others and now share in the nourishing sap from the olive root,

18 do not boast over those branches. If you do, consider this: You do not support the root, but the root supports you.

19 You will say then, "Branches were broken off so that I could be grafted in."

20 Granted. But they were broken off because of unbelief, and you stand by faith. Do not be arrogant, but be afraid.

21 For if God did not spare the natural branches, he will not spare you either.

22 Consider therefore the kindness and sternness of God: sternness to those who fell, but kindness to you, provided that you continue in his kindness. Otherwise, you also will be cut off.

23 And if they do not persist in unbelief, they will be grafted in, for God is able to graft them in again.

24 After all, if you were cut out of an olive tree that is wild by nature, and contrary to nature were grafted into a cultivated olive tree, how much more readily will these, the natural branches, be grafted into their own olive tree!

Romans 13, 4 verses

130

1 Everyone must submit himself to the governing authorities, for there is no authority except that which God has established. The authorities that exist have been established by God.

2 Consequently, he who rebels against the authority is rebelling against what God has instituted, and those who do so will bring judgment on themselves.

3 For rulers hold no terror for those who do right, but for those who do wrong. Do you want to be free from fear of the one in authority? Then do what is right and he will commend you.

4 For he is God's servant to do you good. But if you do wrong, be afraid, for he does not bear the sword for nothing. He is God's servant, an agent of wrath to bring punishment on the wrongdoer.

Romans 14, 3 verses

10 You, then, why do you judge your brother? Or why do you look down on your brother? For we will all stand before God's judgment seat.

11 It is written: "'As surely as I live,' says the Lord, 'every knee will bow before me; every tongue will confess to God.'"

12 So then, each of us will give an account of himself to God.

Romans 15, 2 verses

5 May the God who gives endurance and encouragement give you a spirit of unity among yourselves as you follow Christ Jesus,

6 so that with one heart and mouth you may glorify the God and Father of our Lord Jesus Christ.

1 Corinthians Total: 26 Verses

1 Corinthians 1, 6 verses

20 Where is the wise man? Where is the scholar? Where is the philosopher of this age? Has not God made foolish the wisdom of the world?

21 For since in the wisdom of God the world through its wisdom did not know him, God was pleased through the foolishness of what was preached to save those who believe.

27 But God chose the foolish things of the world to shame the wise; God chose the weak things of the world to shame the strong.

28 He chose the lowly things of this world and the despised things-- and the things that are not-- to nullify the things that are,

29 so that no one may boast before him.

30 It is because of him that you are in Christ Jesus, who has become for us wisdom from God-- that is, our righteousness, holiness and redemption.

1 Corinthians 2, 5 verses

9 However, as it is written: "No eye has seen, no ear has heard, no mind has conceived what God has prepared for those who love him"--

10 but God has revealed it to us by his Spirit. The Spirit searches all things, even the deep things of God.

11 For who among men knows the thoughts of a man except the man's spirit within him? In the same way no one knows the thoughts of God except the Spirit of God.

12 We have not received the spirit of the world but the Spirit who is from God, that we may understand what God has freely given us.

13 This is what we speak, not in words taught us by human wisdom but in words taught by the Spirit, expressing spiritual truths in spiritual words.

1 Corinthians 3, 7 verses

6 I planted the seed, Apollos watered it, but God made it grow.

7 So neither he who plants nor he who waters is anything, but only God, who makes things grow.

8 The man who plants and the man who waters have one purpose, and each will be rewarded according to his own labor.

9 For we are God's fellow workers; you are God's field, God's building.

10 By the grace God has given me, I laid a foundation as an expert builder, and someone else is building on it. But each one should be careful how he builds.

19 For the wisdom of this world is foolishness in God's sight. As it is written: "He catches the wise in their craftiness";

20 and again, "The Lord knows that the thoughts of the wise are futile."

1 Corinthians 4, 3 verses

4 My conscience is clear, but that does not make me innocent. It is the Lord who judges me.

5 Therefore judge nothing before the appointed time; wait till the Lord comes. He will bring to light what is hidden in darkness and will expose the motives of men's hearts. At that time each will receive his praise from God.

9 For it seems to me that God has put us apostles on display at the end of the procession, like men condemned to die in the arena. We have been made a spectacle to the whole universe, to angels as well as to men.

1 Corinthians 5, 1 verses

13 God will judge those outside. "Expel the wicked man from among you."

1 Corinthians 6, 1 verse

14 By his power God raised the Lord from the dead, and he will raise us also.

1 Corinthians 8, 3 verses
3 But the man who loves God is known by God.
5 For even if there are so-called gods, whether in heaven or on earth (as indeed there are many "gods" and many "lords"),
6 yet for us there is but one God, the Father, from whom all things came and for whom we live; and there is but one Lord, Jesus Christ, through whom all things came and through whom we live.

2 Corinthians Total: 20 Verses

2 Corinthians 1, 4 verses
3 Praise be to the God and Father of our Lord Jesus Christ, the Father of compassion and the God of all comfort,
4 who comforts us in all our troubles, so that we can comfort those in any trouble with the comfort we ourselves have received from God.
21 Now it is God who makes both us and you stand firm in Christ. He anointed us,
22 set his seal of ownership on us, and put his Spirit in our hearts as a deposit, guaranteeing what is to come.

2 Corinthians 2, 1 verse
14 But thanks be to God, who always leads us in triumphal procession in Christ and through us spreads everywhere the fragrance of the knowledge of him.

2 Corinthians 3, 2 verses
5 Not that we are competent in ourselves to claim anything for ourselves, but our competence comes from God.
6 He has made us competent as ministers of a new covenant-- not of the letter but of the Spirit; for the letter kills, but the Spirit gives life.

2 Corinthians 5, 6 verses
5 Now it is God who has made us for this very purpose and has given us the Spirit as a deposit, guaranteeing what is to come.
17 Therefore, if anyone is in Christ, he is a new creation; the old has gone, the new has come!
18 All this is from God, who reconciled us to himself through Christ and gave us the ministry of reconciliation:
19 that God was reconciling the world to himself in Christ, not counting men's sins against them. And he has committed to us the message of reconciliation.

20 We are therefore Christ's ambassadors, as though God were making his appeal through us. We implore you on Christ's behalf: Be reconciled to God.

21 God made him who had no sin to be sin for us, so that in him we might become the righteousness of God.

2 Corinthians 7, 4 verses

6 But God, who comforts the downcast, comforted us by the coming of Titus,

7 and not only by his coming but also by the comfort you had given him. He told us about your longing for me, your deep sorrow, your ardent concern for me, so that my joy was greater than ever.

8 Even if I caused you sorrow by my letter, I do not regret it. Though I did regret it-- I see that my letter hurt you, but only for a little while--

9 yet now I am happy, not because you were made sorry, but because your sorrow led you to repentance. For you became sorrowful as God intended and so were not harmed in any way by us.

2 Corinthians 8, 1 verse

16 I thank God, who put into the heart of Titus the same concern I have for you.

2 Corinthians 9, 1 verse

14 And in their prayers for you their hearts will go out to you, because of the surpassing grace God has given you.

2 Corinthians 13, 1 verse

4 For to be sure, he was crucified in weakness, yet he lives by God's power. Likewise, we are weak in him, yet by God's power we will live with him to serve you.

Galatians Total: 14 Verses

Galatians 1, 2 verses

15 But when God, who set me apart from birth and called me by his grace, was pleased

16 to reveal his Son in me so that I might preach him among the Gentiles, I did not consult any man,

Galatians 3, 8 verses

5 Does God give you his Spirit and work miracles among you because you observe the law, or because you believe what you heard?

6 Consider Abraham: "He believed God, and it was credited to him as righteousness."

7 Understand, then, that those who believe are children of Abraham.

8 The Scripture foresaw that God would justify the Gentiles by faith, and announced the gospel in advance to Abraham: "All nations will be blessed through you."

9 So those who have faith are blessed along with Abraham, the man of faith.

16 The promises were spoken to Abraham and to his seed. The Scripture does not say "and to seeds," meaning many people, but "and to your seed," meaning one person, who is Christ.

17 What I mean is this: The law, introduced 430 years later, does not set aside the covenant previously established by God and thus do away with the promise.

18 For if the inheritance depends on the law, then it no longer depends on a promise; but God in his grace gave it to Abraham through a promise.

Galatians 4, 4 verses

4 But when the time had fully come, God sent his Son, born of a woman, born under law,

5 to redeem those under law, that we might receive the full rights of sons.

6 Because you are sons, God sent the Spirit of his Son into our hearts, the Spirit who calls out, "Abba, Father."

7 So you are no longer a slave, but a son; and since you are a son, God has made you also an heir.

Ephesians Total: 18 Verses

Ephesians 1, 11 verses

3 Praise be to the God and Father of our Lord Jesus Christ, who has blessed us in the heavenly realms with every spiritual blessing in Christ.

4 For he chose us in him before the creation of the world to be holy and blameless in his sight. In love

5 he predestined us to be adopted as his sons through Jesus Christ, in accordance with his pleasure and will--

6 to the praise of his glorious grace, which he has freely given us in the One he loves.

7 In him we have redemption through his blood, the forgiveness of sins, in accordance with the riches of God's grace

8 that he lavished on us with all wisdom and understanding.

9 And he made known to us the mystery of his will according to his good pleasure, which he purposed in Christ,

10 to be put into effect when the times will have reached their fulfillment-- to bring all things in heaven and on earth together under one head, even Christ.

11 In him we were also chosen, having been predestined according to the plan of him who works out everything in conformity with the purpose of his will,

22 And God placed all things under his feet and appointed him to be head over everything for the church,

23 which is his body, the fullness of him who fills everything in every way.

Ephesians 2, 7 verses

4 But because of his great love for us, God, who is rich in mercy,

5 made us alive with Christ even when we were dead in transgressions-- it is by grace you have been saved.

6 And God raised us up with Christ and seated us with him in the heavenly realms in Christ Jesus,

7 in order that in the coming ages he might show the incomparable riches of his grace, expressed in his kindness to us in Christ Jesus.

8 For it is by grace you have been saved, through faith-- and this not from yourselves, it is the gift of God--

9 not by works, so that no one can boast.

10 For we are God's workmanship, created in Christ Jesus to do good works, which God prepared in advance for us to do.

Philippians Total: 1 Verse

Philippians 2, 1 verse

13 or it is God who works in you to will and to act according to his good purpose.

Colossians Total: 10 Verses

Colossians 1, 9 verses

12 giving thanks to the Father, who has qualified you to share in the inheritance of the saints in the kingdom of light.

13 For he has rescued us from the dominion of darkness and brought us into the kingdom of the Son he loves,

14 in whom we have redemption, the forgiveness of sins.

15 He is the image of the invisible God, the firstborn over all creation.

16 For by him all things were created: things in heaven and on earth, visible and invisible, whether thrones or powers or rulers or authorities; all things were created by him and for him.

17 He is before all things, and in him all things hold together.

18 And he is the head of the body, the church; he is the beginning and the firstborn from among the dead, so that in everything he might have the supremacy.

19 For God was pleased to have all his fullness dwell in him,
20 and through him to reconcile to himself all things, whether things on earth or things in heaven, by making peace through his blood, shed on the cross.

Colossians 2, 1 verse
19 He has lost connection with the Head, from whom the whole body, supported and held together by its ligaments and sinews, grows as God causes it to grow.

1 Thessalonians Total: 7 Verses

1 Thessalonians 1, 1 verse
4 For we know, brothers loved by God, that he has chosen you,

1 Thessalonians 2, 1 verse
4On the contrary, we speak as men approved by God to be entrusted with the gospel. We are not trying to please men but God, who tests our hearts.

1 Thessalonians 2, 2 verses
16 For the Lord himself will come down from heaven, with a loud command, with the voice of the archangel and with the trumpet call of God, and the dead in Christ will rise first.
17 After that, we who are still alive and are left will be caught up together with them in the clouds to meet the Lord in the air. And so we will be with the Lord forever.

1 Thessalonians 5, 2 verses
9 For God did not appoint us to suffer wrath but to receive salvation through our Lord Jesus Christ.
10 He died for us so that, whether we are awake or asleep, we may live together with him.

2 Thessalonians Total: 7 Verses

2 Thessalonians 2, 7 verses
11 For this reason God sends them a powerful delusion so that they will believe the lie
12 and so that all will be condemned who have not believed the truth but have delighted in wickedness.

13 But we ought always to thank God for you, brothers loved by the Lord, because from the beginning God chose you to be saved through the sanctifying work of the Spirit and through belief in the truth.

14 He called you to this through our gospel, that you might share in the glory of our Lord Jesus Christ.

15 So then, brothers, stand firm and hold to the teachings we passed on to you, whether by word of mouth or by letter.

16 May our Lord Jesus Christ himself and God our Father, who loved us and by his grace gave us eternal encouragement and good hope,

17 encourage your hearts and strengthen you in every good deed and word.

1 Timothy Total: 2 Verses

1 Timothy 1, 1 verses

11 that conforms to the glorious gospel of the blessed God, which he entrusted to me.

1 Timothy 4, 1 verse

3 They forbid people to marry and order them to abstain from certain foods, which God created to be received with thanksgiving by those who believe and who know the truth.

2 Timothy Total: 3 Verses

2 Timothy 1, 2 verses

9 who has saved us and called us to a holy life-- not because of anything we have done but because of his own purpose and grace. This grace was given us in Christ Jesus before the beginning of time,

18 May the Lord grant that he will find mercy from the Lord on that day! You know very well in how many ways he helped me in Ephesus.

2 Timothy 4, 1 verse

17 But the Lord stood at my side and gave me strength, so that through me the message might be fully proclaimed and all the Gentiles might hear it. And I was delivered from the lion's mouth.

18   The Lord will rescue me from every evil attack and will bring me safely to his heavenly kingdom. To him be glory forever and ever. Amen.

Titus Total: 3 Verses

Titus 1, 3 verses
1 Paul, a servant of God and an apostle of Jesus Christ for the faith of God's elect and the knowledge of the truth that leads to godliness--
2 a faith and knowledge resting on the hope of eternal life, which God, who does not lie, promised before the beginning of time,
3 and at his appointed season he brought his word to light through the preaching entrusted to me by the command of God our Savior,

Hebrews Total: 106 Verses

Hebrews 1, 3 verses
1 In the past God spoke to our forefathers through the prophets at many times and in various ways,
2 but in these last days he has spoken to us by his Son, whom he appointed heir of all things, and through whom he made the universe.
3 The Son is the radiance of God's glory and the exact representation of his being, sustaining all things by his powerful word. After he had provided purification for sins, he sat down at the right hand of the Majesty in heaven.

Hebrews 2, 8 verses
1 We must pay more careful attention, therefore, to what we have heard, so that we do not drift away.
2 For if the message spoken by angels was binding, and every violation and disobedience received its just punishment,
3 how shall we escape if we ignore such a great salvation? This salvation, which was first announced by the Lord, was confirmed to us by those who heard him.
4 God also testified to it by signs, wonders and various miracles, and gifts of the Holy Spirit distributed according to his will.
5 It is not to angels that he has subjected the world to come, about which we are speaking.
6 But there is a place where someone has testified: "What is man that you are mindful of him, the son of man that you care for him?
7 You made him a little lower than the angels; you crowned him with glory and honor
8 and put everything under his feet." In putting everything under him, God left nothing that is not subject to him. Yet at present we do not see everything subject to him.

Hebrews 3, 13 verses
7 So, as the Holy Spirit says: "Today, if you hear his voice,

8 do not harden your hearts as you did in the rebellion, during the time of testing in the desert,

9 where your fathers tested and tried me and for forty years saw what I did.

10 That is why I was angry with that generation, and I said, 'Their hearts are always going astray, and they have not known my ways.'

11 So I declared on oath in my anger, 'They shall never enter my rest.'"

12 See to it, brothers, that none of you has a sinful, unbelieving heart that turns away from the living God.

13 But encourage one another daily, as long as it is called Today, so that none of you may be hardened by sin's deceitfulness.

14 We have come to share in Christ if we hold firmly till the end the confidence we had at first.

15 As has just been said: "Today, if you hear his voice, do not harden your hearts as you did in the rebellion."

16 Who were they who heard and rebelled? Were they not all those Moses led out of Egypt?

17 And with whom was he angry for forty years? Was it not with those who sinned, whose bodies fell in the desert?

18 And to whom did God swear that they would never enter his rest if not to those who disobeyed?

19 So we see that they were not able to enter, because of their unbelief.

Hebrews 4, 2 verses

12 For the word of God is living and active. Sharper than any double-edged sword, it penetrates even to dividing soul and spirit, joints and marrow; it judges the thoughts and attitudes of the heart.

13   Nothing in all creation is hidden from God's sight Everything is uncovered and laid bare before the eyes of him to whom we must give account.

Hebrews 6, 7 verses

10 God is not unjust; he will not forget your work and the love you have shown him as you have helped his people and continue to help them.

13 When God made his promise to Abraham, since there was no one greater for him to swear by, he swore by himself,

14 saying, "I will surely bless you and give you many descendants."

15 And so after waiting patiently, Abraham received what was promised.

16 Men swear by someone greater than themselves, and the oath confirms what is said and puts an end to all argument.

17 Because God wanted to make the unchanging nature of his purpose very clear to the heirs of what was promised, he confirmed it with an oath.

18 God did this so that, by two unchangeable things in which it is impossible for God to lie, we who have fled to take hold of the hope offered to us may be greatly encouraged.

Hebrews 8, 5 verses

8 But God found fault with the people and said: "The time is coming, declares the Lord, when I will make a new covenant with the house of Israel and with the house of Judah.

9 It will not be like the covenant I made with their forefathers when I took them by the hand to lead them out of Egypt, because they did not remain faithful to my covenant, and I turned away from them, declares the Lord.

10 This is the covenant I will make with the house of Israel after that time, declares the Lord. I will put my laws in their minds and write them on their hearts. I will be their God, and they will be my people.

11 No longer will a man teach his neighbor, or a man his brother, saying, 'Know the Lord,' because they will all know me, from the least of them to the greatest.

12 For I will forgive their wickedness and will remember their sins no more."

Hebrews 10, 20 verses

7 Then I said, 'Here I am-- it is written about me in the scroll-- I have come to do your will, O God.'"

8 First he said, "Sacrifices and offerings, burnt offerings and sin offerings you did not desire, nor were you pleased with them" (although the law required them to be made).

9 Then he said, "Here I am, I have come to do your will." He sets aside the first to establish the second.

10 And by that will, we have been made holy through the sacrifice of the body of Jesus Christ once for all.

11 Day after day every priest stands and performs his religious duties; again and again he offers the same sacrifices, which can never take away sins.

12 But when this priest had offered for all time one sacrifice for sins, he sat down at the right hand of God.

13 Since that time he waits for his enemies to be made his footstool,

14 because by one sacrifice he has made perfect forever those who are being made holy.

15 The Holy Spirit also testifies to us about this. First he says:

16 "This is the covenant I will make with them after that time, says the Lord. I will put my laws in their hearts, and I will write them on their minds."

17 Then he adds: "Their sins and lawless acts I will remember no more."

18 And where these have been forgiven, there is no longer any sacrifice for sin.

19 Therefore, brothers, since we have confidence to enter the Most Holy Place by the blood of Jesus,

20 by a new and living way opened for us through the curtain, that is, his body,

21 and since we have a great priest over the house of God,

22 let us draw near to God with a sincere heart in full assurance of faith, having our hearts sprinkled to cleanse us from a guilty conscience and having our bodies washed with pure water.

23 Let us hold unswervingly to the hope we profess, for he who promised is faithful.

30 For we know him who said, "It is mine to avenge; I will repay," and again, "The Lord will judge his people."

31 It is a dreadful thing to fall into the hands of the living God.

36 You need to persevere so that when you have done the will of God, you will receive what he has promised.

Hebrews 11, 38 verses

3 By faith we understand that the universe was formed at God's command, so that what is seen was not made out of what was visible.

4 By faith Abel offered God a better sacrifice than Cain did. By faith he was commended as a righteous man, when God spoke well of his offerings. And by faith he still speaks, even though he is dead.

5 By faith Enoch was taken from this life, so that he did not experience death; he could not be found, because God had taken him away. For before he was taken, he was commended as one who pleased God.

6 And without faith it is impossible to please God, because anyone who comes to him must believe that he exists and that he rewards those who earnestly seek him.

7 By faith Noah, when warned about things not yet seen, in holy fear built an ark to save his family. By his faith he condemned the world and became heir of the righteousness that comes by faith.

8 By faith Abraham, when called to go to a place he would later receive as his inheritance, obeyed and went, even though he did not know where he was going.

9 By faith he made his home in the promised land like a stranger in a foreign country; he lived in tents, as did Isaac and Jacob, who were heirs with him of the same promise.

10 For he was looking forward to the city with foundations, whose architect and builder is God.

11 By faith Abraham, even though he was past age-- and Sarah herself was barren-- was enabled to become a father because he considered him faithful who had made the promise.

12 And so from this one man, and he as good as dead, came descendants as numerous as the stars in the sky and as countless as the sand on the seashore.

13 All these people were still living by faith when they died. They did not receive the things promised; they only saw them and welcomed them from a distance. And they admitted that they were aliens and strangers on earth.

14 People who say such things show that they are looking for a country of their own.

15 If they had been thinking of the country they had left, they would have had opportunity to return.

16 Instead, they were longing for a better country-- a heavenly one. Therefore God is not ashamed to be called their God, for he has prepared a city for them.

17 By faith Abraham, when God tested him, offered Isaac as a sacrifice. He who had received the promises was about to sacrifice his one and only son,

18 even though God had said to him, "It is through Isaac that your offspring will be reckoned."

19 Abraham reasoned that God could raise the dead, and figuratively speaking, he did receive Isaac back from death.

20 By faith Isaac blessed Jacob and Esau in regard to their future.

21 By faith Jacob, when he was dying, blessed each of Joseph's sons, and worshiped as he leaned on the top of his staff.

22 By faith Joseph, when his end was near, spoke about the exodus of the Israelites from Egypt and gave instructions about his bones.

23 By faith Moses' parents hid him for three months after he was born, because they saw he was no ordinary child, and they were not afraid of the king's edict.

24 By faith Moses, when he had grown up, refused to be known as the son of Pharaoh's daughter.

25 He chose to be mistreated along with the people of God rather than to enjoy the pleasures of sin for a short time.

26 He regarded disgrace for the sake of Christ as of greater value than the treasures of Egypt, because he was looking ahead to his reward.

27 By faith he left Egypt, not fearing the king's anger; he persevered because he saw him who is invisible.

28 By faith he kept the Passover and the sprinkling of blood, so that the destroyer of the firstborn would not touch the firstborn of Israel.

29 By faith the people passed through the Red Sea as on dry land; but when the Egyptians tried to do so, they were drowned.

30 By faith the walls of Jericho fell, after the people had marched around them for seven days.

31 By faith the prostitute Rahab, because she welcomed the spies, was not killed with those who were disobedient.

32 And what more shall I say? I do not have time to tell about Gideon, Barak, Samson, Jephthah, David, Samuel and the prophets,

33 who through faith conquered kingdoms, administered justice, and gained what was promised; who shut the mouths of lions,

34 quenched the fury of the flames, and escaped the edge of the sword; whose weakness was turned to strength; and who became powerful in battle and routed foreign armies.

35 Women received back their dead, raised to life again. Others were tortured and refused to be released, so that they might gain a better resurrection.

36 Some faced jeers and flogging, while still others were chained and put in prison.

37 They were stoned; they were sawed in two; they were put to death by the sword. They went about in sheepskins and goatskins, destitute, persecuted and mistreated--

38 the world was not worthy of them. They wandered in deserts and mountains, and in caves and holes in the ground.

39 These were all commended for their faith, yet none of them received what had been promised.

40 God had planned something better for us so that only together with us would they be made perfect.

Hebrews 12, 8 verses

5 And you have forgotten that word of encouragement that addresses you as sons: "My son, do not make light of the Lord's discipline, and do not lose heart when he rebukes you,

6 because the Lord disciplines those he loves, and he punishes everyone he accepts as a son."

7 Endure hardship as discipline; God is treating you as sons. For what son is not disciplined by his father?

8 If you are not disciplined (and everyone undergoes discipline), then you are illegitimate children and not true sons.

9 Moreover, we have all had human fathers who disciplined us and we respected them for it. How much more should we submit to the Father of our spirits and live!

10 Our fathers disciplined us for a little while as they thought best; but God disciplines us for our good, that we may share in his holiness.

25 See to it that you do not refuse him who speaks. If they did not escape when they refused him who warned them on earth, how much less will we, if we turn away from him who warns us from heaven?

26 At that time his voice shook the earth, but now he has promised, "Once more I will shake not only the earth but also the heavens."

Hebrews 13, 2 verses

4 Marriage should be honored by all, and the marriage bed kept pure, for God will judge the adulterer and all the sexually immoral.

5 Keep your lives free from the love of money and be content with what you have, because God has said, "Never will I leave you; never will I forsake you."

James Total: 22 Verses

James 1, 12 verses

5 If any of you lacks wisdom, he should ask God, who gives generously to all without finding fault, and it will be given to him.

12 Blessed is the man who perseveres under trial, because when he has stood the test, he will receive the crown of life that God has promised to those who love him.

13 When tempted, no one should say, "God is tempting me." For God cannot be tempted by evil, nor does he tempt anyone;

14 but each one is tempted when, by his own evil desire, he is dragged away and enticed.

15 Then, after desire has conceived, it gives birth to sin; and sin, when it is full-grown, gives birth to death.

16 Don't be deceived, my dear brothers.

17 Every good and perfect gift is from above, coming down from the Father of the heavenly lights, who does not change like shifting shadows.

18 He chose to give us birth through the word of truth, that we might be a kind of first fruits of all he created.

19 My dear brothers, take note of this: Everyone should be quick to listen, slow to speak and slow to become angry,

20 for man's anger does not bring about the righteous life that God desires.

21 Therefore, get rid of all moral filth and the evil that is so prevalent and humbly accept the word planted in you, which can save you.

27 Religion that God our Father accepts as pure and faultless is this: to look after orphans and widows in their distress and to keep oneself from being polluted by the world.

James 2, 2 verses

5 Listen, my dear brothers: Has not God chosen those who are poor in the eyes of the world to be rich in faith and to inherit the kingdom he promised those who love him?

23 And the scripture was fulfilled that says, "Abraham believed God, and it was credited to him as righteousness," and he was called God's friend.

James 4, 7 verses

6 But he gives us more grace. That is why Scripture says: "God opposes the proud but gives grace to the humble."

7 Submit yourselves, then, to God. Resist the devil, and he will flee from you.

8 Come near to God and he will come near to you. Wash your hands, you sinners, and purify your hearts, you double-minded.

9 Grieve, mourn and wail. Change your laughter to mourning and your joy to gloom.

10 Humble yourselves before the Lord, and he will lift you up.

11 Brothers, do not slander one another. Anyone who speaks against his brother or judges him speaks against the law and judges it. When you judge the law, you are not keeping it, but sitting in judgment on it.

12 There is only one Lawgiver and Judge, the one who is able to save and destroy. But you-- who are you to judge your neighbor?

James 5, 1 verse

11 As you know, we consider blessed those who have persevered. You have heard of Job's perseverance and have seen what the Lord finally brought about. The Lord is full of compassion and mercy.

1 Peter Total: 9 Verses

1 Peter 1, 6 verses

3 Praise be to the God and Father of our Lord Jesus Christ! In his great mercy he has given us new birth into a living hope through the resurrection of Jesus Christ from the dead,

4 and into an inheritance that can never perish, spoil or fade-- kept in heaven for you,

5 who through faith are shielded by God's power until the coming of the salvation that is ready to be revealed in the last time.

17 Since you call on a Father who judges each man's work impartially, live your lives as strangers here in reverent fear.

20 He was chosen before the creation of the world, but was revealed in these last times for your sake.

21 Through him you believe in God, who raised him from the dead and glorified him, and so your faith and hope are in God.

1 Peter 3, 1 verse

12 For the eyes of the Lord are on the righteous and his ears are attentive to their prayer, but the face of the Lord is against those who do evil.

1 Peter 5, 2 verses

6 Humble yourselves, therefore, under God's mighty hand, that he may lift you up in due time.

7 Cast all your anxiety on him because he cares for you.

2 Peter Total: 14 Verses

2 Peter 1, 5 verses

16 We did not follow cleverly invented stories when we told you about the power and coming of our Lord Jesus Christ, but we were eyewitnesses of his majesty.

17 For he received honor and glory from God the Father when the voice came to him from the Majestic Glory, saying, "This is my Son, whom I love; with him I am well pleased."

18 We ourselves heard this voice that came from heaven when we were with him on the sacred mountain.

20 Above all, you must understand that no prophecy of Scripture came about by the prophet's own interpretation.

21 For prophecy never had its origin in the will of man, but men spoke from God as they were carried along by the Holy Spirit.

2 Peter 2, 6 verses

4 For if God did not spare angels when they sinned, but sent them to hell, putting them into gloomy dungeons to be held for judgment;

5 if he did not spare the ancient world when he brought the flood on its ungodly people, but protected Noah, a preacher of righteousness, and seven others;

6 if he condemned the cities of Sodom and Gomorrah by burning them to ashes, and made them an example of what is going to happen to the ungodly;

7 and if he rescued Lot, a righteous man, who was distressed by the filthy lives of lawless men

8 (for that righteous man, living among them day after day, was tormented in his righteous soul by the lawless deeds he saw and heard)--

9 if this is so, then the Lord knows how to rescue godly men from trials and to hold the unrighteous for the day of judgment, while continuing their punishment.

2 Peter 3, 3 verses

5 But they deliberately forget that long ago by God's word the heavens existed and the earth was formed out of water and by water.

6 By these waters also the world of that time was deluged and destroyed.

7 By the same word the present heavens and earth are reserved for fire, being kept for the day of judgment and destruction of ungodly men.

1 John Total: 23 Verses

1 John 1, 1 verse

3 We proclaim to you what we have seen and heard, so that you also may have fellowship with us. And our fellowship is with the Father and with his Son, Jesus Christ.

1 John 3, 3 verses

1 How great is the love the Father has lavished on us, that we should be called children of God! And that is what we are! The reason the world does not know us is that it did not know him.

5 But you know that he appeared so that he might take away our sins. And in him is no sin.

24 Those who obey his commands live in him, and he in them. And this is how we know that he lives in us: We know it by the Spirit he gave us.

1 John 4, 12 verses

6 We are from God, and whoever knows God listens to us; but whoever is not from God does not listen to us. This is how we recognize the Spirit of truth and the spirit of falsehood.

7 Dear friends, let us love one another, for love comes from God. Everyone who loves has been born of God and knows God.

8 Whoever does not love does not know God, because God is love.

9 This is how God showed his love among us: He sent his one and only Son into the world that we might live through him.

10 This is love: not that we loved God, but that he loved us and sent his Son as an atoning sacrifice for our sins.

11 Dear friends, since God so loved us, we also ought to love one another.

12 No one has ever seen God; but if we love one another, God lives in us and his love is made complete in us.

13 We know that we live in him and he in us, because he has given us of his Spirit.

14 And we have seen and testify that the Father has sent his Son to be the Savior of the world.

15 If anyone acknowledges that Jesus is the Son of God, God lives in him and he in God.

16 And so we know and rely on the love God has for us. God is love. Whoever lives in love lives in God, and God in him.

17 In this way, love is made complete among us so that we will have confidence on the day of judgment, because in this world we are like him.

148

1 John 5, 7 verses

14 This is the confidence we have in approaching God: that if we ask anything according to his will, he hears us.

15 And if we know that he hears us-- whatever we ask-- we know that we have what we asked of him.

16 If anyone sees his brother commit a sin that does not lead to death, he should pray and God will give him life. I refer to those whose sin does not lead to death. There is a sin that leads to death. I am not saying that he should pray about that.

17 All wrongdoing is sin, and there is sin that does not lead to death.

18 We know that anyone born of God does not continue to sin; the one who was born of God keeps him safe, and the evil one cannot harm him.

19 We know that we are children of God, and that the whole world is under the control of the evil one.

20 We know also that the Son of God has come and has given us understanding, so that we may know him who is true. And we are

2 John Total: 4 Verses

2 John 1, 4 verses

3 Grace, mercy and peace from God the Father and from Jesus Christ, the Father's Son, will be with us in truth and love.

4 It has given me great joy to find some of your children walking in the truth, just as the Father commanded us.

5 And now, dear lady, I am not writing you a new command but one we have had from the beginning. I ask that we love one another.

6 And this is love: that we walk in obedience to his commands. As you have heard from the beginning, his command is that you walk in love.

Chapter 4
Scientific Evidence of Creation

For since the creation of the world God's invisible qualities-- his eternal power and divine nature-- have been clearly seen, being understood from what has been made, so that men are without excuse. (Rom 1:20, NIV)

Enchanted Forest, Mount Hood, Oregon.

We all have a worldview that influences our perception of the world around us. A scientist with an atheistic worldview sees undirected evolution as the explanation for the origin, diversity and complexity of life. Secular scientists see the fossil record as evidence of change over vast periods of time. They reject the Biblical teaching of the global flood and its impact on the world. In sharp contrast, the scientist with a Christian worldview sees the complexity, diversity and beauty of living things as the result of the Creator. Even the placement of the earth in our galaxy is seen as optimal for human existence against odds that seem most unlikely. They see the fossil record is scientific proof of the global flood described in the Bible and confirmed by the Jesus Christ Himself as an example of His return. *For in the days before the flood, people were eating and drinking, marrying and giving in marriage, up to the day Noah entered the ark; and they knew nothing about what would happen until the flood came and took them all away. That is how it will be at the coming of the Son of Man* (Matt 24:38-39). The original world was created without suffering, disease, or death, but due to Adam and Eve's sin, that world was forever changed. We now live under the curse and can only imagine what life was like in the Garden of Eden. *We know that the whole creation has been groaning as in the pains of childbirth right up to the present time* (Rom 8:22, NIV). Christians will see the return of that original perfect world in the new heaven and new earth (Rev 21:1). A person looking at the complexity of living things with an open mind can not help but see the overwhelming evidence for Intelligent Design leading to the Creator of all things.

Naturalistic evolution has failed to provide a plausible explanation for the origin and complexity of life. Mutations have long been seen as harmful errors that reduce genetic information. They are never creative and cannot account for the complexity and variety in living things. Macroevolution lacks convincing scientific support or even a demonstrated mechanism and is rapidly falling into disrepute. Evolutionists do not have a plausible theory to account for the origin of life or its complexity. With the exiting new discoveries about the extreme complexity of all living things, the very foundation of evolution is beginning to crack and many scientists believe the entire edifice will soon crumble. Evolution is bankrupt. Thousands of former evolutionists no longer see evolution as having a reasonable explanation for the origin of life or its complexity and are abandoning evolution like rats from a sinking ship. The Psalmist said it best. *The heavens declare the glory of God; the skies proclaim the work of his hands. Day after day, they pour forth speech; night after night, they display knowledge. There is no speech or language where their voice is not heard* (Ps 19:1-3, NIV). Modern science is hearing this speech.

Another important point in that insightful passage needs to be emphasized. Those failing to recognize the God of Creation are without excuse. God loves us so much that He made the Creation in such a way that declares His power and glory for *all* to see. The miracle of Creation has not been hidden from view. *For since the creation of the world God's invisible qualities-- his eternal power and divine nature-- have been clearly seen, being understood from what has been made, so that men are without excuse. For although they knew God, they neither glorified him as God nor gave thanks to him, but their thinking became futile and their foolish hearts were darkened. Although they claimed to be wise, they became fools* (Rom 1:20-22, NIV). Anyone willing to look objectively at nature sees the overwhelming evidence of God as Creator. The Apostle Peter says this blindness to creation and to the global flood is due to **willful ignorance** of sinful man (2 Pet. 3:3-7, KJV).

Why then do the secular scientists fail to see the obvious evidence for Creation? The answer is simple: they *must* deny the evidence in order to cling to their atheistic worldview. ***Biologists must constantly keep in mind that what they see was not designed, but rather evolved*** (Crick, 1986). Notice the secular biologist "must constantly keep in mind" that macroevolution brought every living thing into the world and not the self-evident God of creation. Evolutionists accuse Christians who attribute the origin of life to God as unscientific myth lacking merit. In order to do this, they must ignore the historical fact that many of the great scientists of the past saw no conflict with science and the concept of God as Creator. When it comes to relying on miracles, evolutionists are just as guilty as Christians! ***An honest man, armed with all the knowledge available to us now, could only state that in some sense, the origin of life appears at the moment to be almost a miracle, so many are the conditions which would have to be satisfied to get it going*** (Crick, 1981). The evolutionists have the greater measure of blind faith for Christians know in whom they believe. ***Yet I am not ashamed, because I know whom I have believed, and am convinced that he is able to guard what I have entrusted to him for that day*** (2 Tim 1:12b, NIV).

In this chapter, we will consider some of the scientific evidence supporting supernatural creation and see why the facts of science are detrimental to materialistic macroevolution. I am not speaking as an outsider, a preacher or as a philosopher, but as a well published zoologist. I understand first-hand how science operates. I spent most of my professional life investigating thermoregulation in alligators and the cardiovascular response of wild animals to fear. In addition, I designed and published the radio telemetry devices that made those studies possible. I have implanted heart rate transmitters in more species of wild animals than anyone in the world. I have lectured at graduate and medical schools on three continents and have published nearly over one hundred peer reviewed technical papers, abstracts, magazine articles and books.

Following are a few examples of the scientific evidences for creation. Each of these is a veritable nightmare for evolutionists. Many other examples could be given and more are being discovered each year. Perhaps the most devastating recent evidence comes from molecular biology and that will be discussed in the chapter dealing with Intelligent Design. When a new discovery is made that appears to support evolution, it is widely publicized, yet we never hear of the countless discoveries that are devastating to evolution dogma and support Creation. Discoveries that cannot be explained by evolution are not addressed in textbooks or mentioned by professors in university classrooms. They are swept under the carpet. Such important findings must be understood and openly addressed. We must understand and be willing and able to share the

153

overwhelming scientific evidence supporting the supernatural Creation. We have the facts on our side. Let us begin our journey on the road less traveled.

Scientific Evidence for Creation From Astronomy

Astronomy is the oldest of the natural sciences and dates back to antiquity. Early astronomy involved human observations with the naked eye of the regular and predictable patterns of motion of the sun, moon, stars and larger planets. It also has mythological origins including astrological practices. The movements of the stars and planets were to be used as signs of the seasons and were important for agriculture and other activities as indicated in Genesis. ***And God said, "Let there be lights in the expanse of the sky to separate the day from the night, and let them serve as signs to mark seasons and days and years.*** (Gen 1:14) This was their God-ordained purpose.

Early calendars were established in most cultures by the movement of the moon and sun and were used to indicate weeks, months and years. Our modern calendar including the necessity of leap year was developed 400 years before Christ. It was known the stars remained fixed, while the planets moved freely and predictably. Solar and lunar eclipses have been predicted for thousands of years and even the earth's diameter was measured from eclipse measurements 1,500 years ago. The Copernican revolution marked the marriage of science and astronomy with the proposal by Nicolaus Copernicus (1473-1543) that the planets including the earth revolved around the sun. By the 1700's astronomy became an important and permanent part of physics.

Astronomy was forever changed on July 20, 1969 when Astronaut Neil Armstrong stepped onto the moon and uttered those unforgettable words, "That's one small step for man, one giant leap for mankind." A new age for astronomy began. An estimated 500 million people witnessed the moon landing, the largest television audience for a single event in history. In the intervening years, our knowledge of astronomy has exploded. The Hubble Telescope, launched in 1990, provides breathtakingly beautiful photographs and measurements from deep space never before possible and greatly expanded our knowledge of the size and complexity of the universe. The Mars explorer revealed secrets from the Red Planet and countless other space probes have provided a treasure-trove of data from beyond our planet.

Throughout the twentieth century, it was generally accepted there was nothing special about the earth. Astronomers discovered there were billions of galaxies, each with billions of stars and untold billions of planets. Our earth was considered a mere speck of dust in the vastness of space. It was assumed there were billions of other earth-like planets and many of them had intelligent life more advanced than our own. Such was the motivation for SETI (Search for Extraterrestrial Intelligence) in the 1960's. For decades, powerful radio

Spiral Galaxy M74 from the Hubble Telescope.

telescopes searched the galaxies for signs of intelligent alien life, but found nothing. It is astounding that this search for intelligent design is accepted by secular science while recent attempts to explain the complexity of living cells as evidence of intelligent design have been patently rejected as "unscientific" or "religiously motivated."

Today there is a growing appreciation for the uniqueness of planet earth. Perhaps no one has had a greater impact on this important sea change than astrobiologist Guillermo Gonzalez. In his popular book and subsequent movie, **The Privileged Planet,** Dr. Gonzalez listed twenty characteristics necessary for intelligent life. All of them must be present for life as we know it to exist. In seems the earth is special after all and perhaps even unique. Earth-like planets with all the requirements to sustain life are now considered quite rare. As we learn more about our planet and its place in the universe, the evidence the earth was designed for man is overwhelming. One can apply simple probability theory to calculate the odds for a life supporting planet. The results are truly staggering. In a conservative estimate of the probability of earth-like conditions necessary to support complex life, researchers found the phenomenal odds against such a planet as being 1 to 1,000,000,000,000,000. It is not a good time to be an evolutionist. Pesky facts like this keep getting in the way of good dogma.

Perhaps the strongest proof of creation from astronomy was the most famous of all stars, the Star of Bethlehem announcing the birth of Jesus Christ. Using NASA charts showing the precise position of the stars and planets throughout history, it is possible to describe this striking event matching all twelve characteristics described in the Bible to be the precise juxtaposition of the largest planet alongside the largest star and it has only occurred once, precisely in time to announce the birth of the Messiah. From a careful examination of history and the new information from NASA, we now know each of the nine things mentioned in scripture about this special heavenly body occurred at the exact time of the birth of Christ. The list includes: the star signified a birth, it signified kingship, it had a connection with the Jewish nation, it rose in the east, like other stars, it appeared at a precise time in history, King Herod did not know when it appeared, it endured over time, it was ahead of the Magi as they went south from Jerusalem to Bethlehem and it stopped over Bethlehem. There was also a prophesied corresponding solar eclipse at the time Jesus was crucified. For details about these remarkable events see www.bethlehemstar.net.

Scientific Evidence for Creation From Zoology

As a lifelong zoologist, I find these and many more such evidences exciting for the support a supernatural Creation and provide strong evidence against evolution by random chance. Following are a few of many evidences support creation from zoology that present insurmountable problems for evolutionists. Some are from my own research and musings about the wonders of Creation. Others are known and published, but all are avoided by evolutionists for good reason as we shall see.

To Kill a Bison and Feed a People

It is difficult to imagine an indigenous people more closely connected to an animal than were our own Native Americans to the American bison or buffalo. Even today, many tribal leaders mark time from before and after the disappearance buffalo. It is estimated over 60 million bison once roamed the vast western prairies from Mexico to Canada. Their slaughter by white hunters for pelts and later only for their tongues brought near extinction to bison and marked the end of a way of life that had been in harmony with nature for thousands of years. The native people depended on the virtually unlimited bison for food, clothes, tools, medicine, ornaments and shelter. Have you ever wondered at the success of Native Americans in killing this huge animal? For decades, I marveled at how the native people could have been so successful in harvesting this important and magnificent creature. After considerable research, I now understand and the reason points clearly to the Creator and is impossible for evolutionists to explain.

Historically, there were four important ways bison were killed by Native American people. Some were stampeded over cliffs or trapped in box canyons. Both of these methods were effective, but could only be used in a limited number of places and then only if the bison herd was at the right place. Neither method could reliably support the plains people for centuries.

In the sixteenth century, Spaniards and others re-introduced horses to America. It is thought over-hunting caused their earlier extinction toward the end of the ice age. Their influence on native people was immediate and profound. A third method of killing bison was running down a single animal by horse. Bison are powerful animals and it often took as many as five or more fresh horses to finally fatigue the bison.

Horses were more commonly used to kill bison as often seen in movies. A horseback rider would run alongside a running bison and shoot it with a bow and arrow. This method troubled me for years. I am not a bow hunter, but the idea of hitting the heart of a running bison from a galloping horse seemed difficult if not impossible. The answer came in a graduate course in vertebrate natural history at Baylor University. This problem was uniquely solved in a surprising way by the Creator of man and bison.

American Bison, An Evolutionist's Nightmare.

With the notable exception of the American bison, *all* mammals have two separate lung or pleural cavities. As we know, one side of our chest can be penetrated collapsing that lung, but the other side remains intact and that lung can support life. The bison is unique in having an incompletely divided mediastinum. There is only one functional pleural cavity containing both lungs. Thus the problem for the native bow hunter is solved. An arrow must only penetrate the chest and both lungs collapse. The fatally wounded animal would continue a few yards and die providing unlimited food, clothing and tools. Before the availability of horses, bison could be shot by stealth from a blind or other hiding place. One problem is solved, yet another serious one remains. This problem is *never* mentioned in biology or evolution classes, yet this important problem demands an answer.

The problem for the evolutionist is simple, yet revealing and difficult. Other than providing food for hungry people, of what possible selective advantage is an incompletely divided mediastinum? From an evolutionary viewpoint, it makes absolutely no sense and does *not* occur in any other mammal. Indeed, conventional wisdom would argue for the elimination of such a detrimental trait from the gene pool. Yet it remained, providing food, clothes and tools for a continent of Native Americans for millennia. It must indeed require faith and dedication to remain an evolutionist for the facts of science keep getting in the way. I know the Creator of the bison and understand this profound demonstration of love and provision for Native American people.

God's Pregnancy Test

Long before the corner drug store had inexpensive pregnancy test kits available, before the rabbit test was widely used, God provided an easy-to-use absolutely free pregnancy test for women. I shared this with my nursing students and was shocked not a single female student was aware of this simple test. As a young man, when I first learned about a woman's reproduction cycle, I instantly put it together with my knowledge of the lunar cycle and was in awe of the Creator for making such important information readily available to all women.
In simpler times before the Internet, cell phones, television and electric lights, people were closely tied to natural light and the phases of the moon. God's Word clearly states: ***And God said, Let there be lights in the expanse of the sky to separate the day from the night, and let them serve as signs to mark seasons and days and years*** (Gen 1:14, NIV). Early civilizations were aware of this and marked such celestial events as spring and fall equinox and the longest and shortest days. Long before the advent of months, years, and even before calendars were invented, a woman could simply go outside at night, look at the moon and determine if she might be pregnant.

How could they do this? It is really quite simple. God designed the lunar cycle to be roughly twenty-nine days in length. This is the period from new moon to new moon or full moon to full moon and is amazingly close to the average woman's 28 day menstrual cycle. Here is how it works. If a woman started her period last month with a full moon and the next full moon came without a period she knew she immediately that she might pregnant. How could it be simpler? Certainly this is not 100 percent accurate, for the menstrual cycle can vary, but it still provided women with an early indication of possible pregnancy.

This raises yet another problem for evolutionists. Why would the lunar cycle agree so closely with a woman's menstrual cycle? Did people evolve so the moon could be used as a pregnancy test? Not likely! Or did the two events just happen to coincide? Again most unlikely; figure the odds! It is far more logical that the God who hung the moon and stars in place and created man in His image planned the cycle of the moon so it could be used as a "sign" of possible pregnancy. Today we think of the moon as being for lovers. So did our Creator. The phase of moon has been used since the beginning of Creation to tell a woman if she was with child. *The heavens declare the glory of God; the skies proclaim the work of his hands* (Ps 19:1). Sadly, such truths can no longer be taught in our public schools or universities. Our teachers are strictly forbidden to mention any evidence from science that points to creation or to God. Instead, they are required to teach materialistic evolution and that man is nothing more than "matter in motion." This must change. Teachers, like scientists should be allowed to follow the evidence, no matter where it leads and a great deal of scientific evidence leads directly to the God of Creation.

Sex Determination in Alligators and Turtles

Here are some very difficult problems for evolutionists that few people know about. Because they present insurmountable problems for evolution, they are not mentioned in textbooks or discussed in classrooms. The problems are related to sex determination in alligators and turtles. First, let's look at some background.

Alligators must be eight to ten years old and approximately six feet long to breed. Most mature females breed three out of every four years. They construct a large nest of nearby vegetation and deposit an average of forty-seven eggs in a depression near the top of the nest. After the eggs are laid, she covers them with more vegetation. Alligator eggs require nine to twelve weeks to hatch and are incubated by the heat from the sun and heat produced by of the decaying vegetation. During the time the eggs are incubating, the female alligator has a restricted home range of less than a third of an acre and most will actively defend the nest. Her presence deters most natural enemies.

Female Alligator Defending Her Nest.

A few hours before hatching, baby alligators inside the egg begin making a grunting sound which synchronizes the hatching so all the baby alligators emerge from the eggs within a few hours. Upon hearing the sound of her emerging young, the mother alligator carefully removes the nest material from the eggs to release the hatchlings. If the mother has been killed, the young alligators remain imprisoned inside the nest and die. Mother alligators have been seen gently breaking the egg shell with their teeth when a baby alligator has difficulty hatching. The female will take each baby alligator into her mouth and carry them to the water where she releases them and returns to the nest for another load. In areas where alligators winter in underground burrows the young stay with her the first two years. Such maternal care is uncommon among reptiles and is largely unknown by evolutionists. They teach maternal instincts did not evolve until the appearance of birds. God was not restricted by evolution dogma and made all living things optimally adapted for their environment. ***God saw all that he had made, and it was very good*** (Gen 1:31a, NIV).

It is the incubation temperature and not sex chromosomes that determine the sex of baby alligators. This presents an even more difficult problem for evolutionists. Alligator reproduction is well known and often shown on television nature programs. Alligators and certain other reptiles lack sex chromosomes (Ferguson and Joanen, 1982). Ted Joanen and Larry McNease of the Rockefeller Wildlife Refuge in Grand Chenier, Louisiana taught me how to call, capture and handle alligators safely and I have returned several times. Dr. Ruth Elsie and others continue alligator research at the refuge. The refuge has a higher

160

Alligator Eggs in the Nest.

concentration of alligators than any other place in the world and is a beautiful place to visit, watch birds and observe alligators in their natural environment.

Research has shown that an average incubation temperature of 85 degrees Fahrenheit yields only female hatchlings. At a temperature of 89 degrees, equal numbers of male and female hatchlings result. At 91 degrees, only males are produced. The investigators admitted in the original paper that, "There has been no demonstration of a selective evolutionary advantage of the occurrence of temperature sex determination in reptiles." One is hard pressed to present an argument in favor of this sort of sex determination as opposed to the more common method of sex chromosomes. If there is a selective advantage, then why do most animals rely on sex chromosomes?

These are difficult problems for evolutionists, but there is an even worse problem. If one could conjure up a selective advantage for such results in crocodilians, the argument runs amuck with the consideration of turtles. Most turtles also lack sex chromosomes and the sex of the offspring is determined by the average incubation temperature of the eggs, but there is a remarkable difference. In turtles, higher temperatures produce females, not males! The notable exception is *Trionyx spiniferus* the spiny soft-shelled turtle which have sex chromosomes and the sex of the hatchlings is unaffected by egg incubation temperature (Bull and Vogt, 1979).

This leaves the evolutionist with an unenviable situation. Whatever arguments are fabricated to support temperature induced sex determination in crocodilians disintegrate when one considers turtles. It is not surprising evolutionists are unwilling to address these important problems for they know they are standing on shifting sand. Could this whole sex-in-response-to-temperature thing have been designed by a Creator with a sense of humor? It seems this would require less faith to accept than some yet-to-be-conceived

161

alleged advantage of allowing temperature to determine sex one way in crocodilians and the opposite way in most, but not all turtles. Once again it is not a good time to be an evolutionist. Facts like these continue unrelenting to give them nightmares.

Lung Bypass Valve

Alligators and other crocodilians present yet another enigma for evolutionists. Their hearts have a special feature that helps them hold their breath and remain submerged for an extended time. It is called the foramen of Panizza and allows blood to bypass the lungs while submerged for prolonged times. It functions much like the foramen ovale in reducing the work load of the heart for unborn mammals by providing a right to left path for blood inside the heart, bypassing the lungs. With the first breath of air the pulmonary resistance to blood flow drops and the valve closes. It quickly grows closed in mammals, but the special feature of crocodilians remains functional throughout their life and helps them occupy their unique ecological niche by allowing them to remain submerged where they are safe for hours. What is particularly troublesome for evolutionists is its uniqueness to crocodilians. Other diving animals such as turtles, birds or mammals lack this special feature. Evolutionists are once again at a loss to explain its origin in crocodilians and the lack of such a useful device in other diving animals. With the avalanche of such new information, it is becoming increasingly difficult to remain a committed evolutionist! The facts continue to stand in the way of good evolution dogma. The cracks in the foundation of evolution are widening each year as we discover more about the complexity, beauty and sheer wonder of God's creation. It requires a huge amount of blind faith to remain a devoted evolutionist. As Christians we know in Whom we believe.

To Run Or Not To Run

Although I was the first to specifically investigate the heart rate response of wild animals to fear, other investigators made similar discoveries by accident while studying something else. I discovered when wild animals are frightened by the approach of man, dog or other predator they often hide, if a safe hiding place is available. While hiding, they remain motionless and reduce their heart and respiration rates. In contrast, if a safe hiding place is not present, they ran with the classic flight or fight response and increased heart and respiration rates. In other words, frightened animals hide if they can, but will run or fight if a safe hiding place is unavailable. Their cardiovascular response is opposite depending upon the response exhibited. These results have now been repeated by many other scientists with a wide variety of animals and situations.

162

White-tailed Deer Fawn.

An interesting age related change in the behavior and heart rate response to fear of deer fawn was reported by Jacobsen (1979). Young fawns are incapable of out-running common predators such as foxes and coyotes. When frightened by the approach of a predator or man, they drop and hide showing a reduction in both heart rate and breathing depth. Hiding often saves their lives.

After the age of about two weeks, profound behavioral and physiologic changes occur in the fawns. They can now outrun predators. In doing so, they have the typical fight or flight response and exhibit increased heart and respiration rates. The adaptive value is obvious, for they can now elude a predator by fleeing, but a problem remains for the evolutionist. How could such a profoundly different response in both behavior and physiology evolve by small steps? Remember, according to evolution each change must be small in order to be inherited, but each step must also be adaptive or it will be quickly eliminated. Once again, it is difficult to remain an evolutionist when one looks objectively at the actual evidence from science.

Kangaroo Rats; Multiple Nightmares for Evolutionists
While walking one morning at sunrise in the beautiful desert of southern Arizona, I realized how truly underrated kangaroo rats are and what a problem they are for evolutionists. They occur on my farm in western Oklahoma, but here

they were abundant with burrows and fresh tracks everywhere. Kangaroo rats are beautiful little brown and white animals with huge liquid brown eyes and a perky gait akin to kangaroos. As we will see, they have given us so much, yet we have returned little. Let me explain.

Ord's kangaroo rat, *Dipodomys ordii* www.biotopics.co.uk

One of their remarkable features is the number of facial muscles, yet this is never discussed in biology classrooms, partly I think because if flies in the face of established evolutionary thinking. Before the twentieth century it was widely publicized that of all mammals, primates had the most facial muscles. The large number of facial muscles is necessary for expression and is used in non-verbal communication. This remains widely accepted, in spite of well known fact that an elephant's trunk possesses over a thousand individual muscles. Primates do in fact have over twice the number of facial muscles of most other mammals. Emotions can be read by changing facial expressions. We see no corresponding facial grimaces in other animals such as livestock, dogs or cats. It was concluded that such communication was correlated with social behavior and intelligence and thus the maximum development and diversity of facial muscles were seen in the most intelligent animals. Man is always seen as the glorious climax of evolution. Such simplistic thinking seldom withstands the test of time.

Remember this is the same crowd that for decades taught the human brain was the largest and most highly evolved of all animal brains. This was until scientists noticed the brains of whales and even the pelvic brain of certain dinosaurs (an enlargement that controls the hind legs) was larger than human brains. Evolutionists quickly modified the definition to say the human has the largest brain as percentage of body mass and thus is proof man is the most highly evolved of all creatures. Once again, there were problems as other animals were soon found to have larger brains even as a percent of body mass. The theory was once again modified to state the human brain is the most convoluted and thus has the greatest surface area of any animal. Evolutionists have not yet recovered from the discovery that whales and porpoises have both relatively larger and more convoluted brains than humans. Again, those troublesome facts keep getting in the way of good evolution legend.

Let's return to our discussion of kangaroo rats. Several years ago, a graduate student discovered kangaroo rats have a surprising forty-two pairs of facial muscles! So much for the theory that intelligence is correlated with the number of facial muscles. Such an obvious question demands a thoughtful answer. Of what possible reason could a small rodent use with so many facial muscles? Even the casual observer realizes kangaroo rats do not appear to spend much time making faces at one another. Here the unique kangaroo rat's gait comes into play. They are often bipedal...hopping like their larger Australian namesakes. But they do not hop very high. Night photography revealed a striking discovery. As they hop, their vibrissa or sensitive "whiskers" maintain contact with the surface! Their need for such a large number facial muscles is for the purpose of providing detailed tactile information about their environment as they hop about. Of course, nothing is written about how such a complex dynamic surface scanning system and its extraordinarily complex interpretation might have evolved. The insignificant kangaroo rat puts yet another fly in the evolutionist's ointment, but there is more.

Kangaroo rats make excellent pets. I kept several as a kid growing up on a farm. This was before they were protected by law and it was legal to keep such creatures. They require little care and are delightful to watch. They eat little and require absolutely no drinking water. Unlike most other mammals, they are virtually odor free. Their lack of need for drinking water is perhaps their most outstanding physiological feature. For years it was observed that many desert inhabitants require little or no water. Scientists have worked out daily water budgets for a wide variety of animals including desert creatures. They carefully measured all water intakes and collected urine and feces to determine the amount of water lost. Respiratory and skin losses of water were also measured. Perplexed by what they found, some biologists postulated dew as a source of water. Obviously those biologists had not spent much time in the desert for dew is a rarity and any animal depending on dew would not survive.

Other biologists suggested the importance of "pre-formed" water found naturally in the desert. Even certain desert plants contain high concentrations of water. Careful observation however clearly showed kangaroo rats could live without such plants. They can subsist entirely on a diet of dry seeds. This means they thrive on the metabolic water produced when carbohydrates are used producing carbon dioxide and water. We all remember from grade school that when sugar is metabolized the end products are carbon dioxide and water. This is the only water required by these delightful little creatures.

Examination of their urine was another biological bombshell. They can produce urine three times more salty than seawater! This observation eventually led to a better understanding of how humans produce urine. The most important functional part of the kidney is the loop of Henley. This is the portion that concentrates urine and saves water. It was discovered the loop of Henley of kangaroo rats was proportionally nearly three times longer than it is in the human. Farther studies showed the primary function of the loop of Henley to be water absorption and retention. Thus understanding kidney function in kangaroo rats led directly to understanding of how the human Loop of Henley functions.
Such is often the case. The human body can do lots of things (physiologically speaking) but cannot do any of them particularly well. It is when we study extreme animal forms, those living on the physiological edge so to speak, that we better understand human function. Rattlesnakes can live up to two years without food or water and parasitic ticks can survive five years with neither blood nor water! No doubt they, too, have secrets we will someday unlock. See why I love zoology? See how clearly our Creator reveals Himself in nature?

It is because the kangaroo rat can produce urine three times more salty that they can survive in the desert southwest without the need to drink water. Humans can only produce urine slightly more salty than seawater. We have all seen movies of those lost at sea and some might wonder then why we cannot survive by drinking seawater. The answer is that the third most abundant salt in seawater is magnesium, which causes excessive water loss from diarrhea when sea water is ingested in quantity. We have learned much from the kangaroo rat and no doubt more remains to be learned. What do you think about while walking in the desert? I think of kangaroo rats and their all-wise Creator and find myself lifting my hands in joyous praise.

Fruit Fly, A Genetics Workhorse

Since the early 1900's, the biological community has used the tiny fruit fly, *Drosophila melanogaster*, to conduct countless experiments. Students in biology classes work with fruit flies, crossing various types to produce predictable inheritance patterns and counting the results. There are tens of thousands of publications dealing with decades of genetic research with fruit flies. To secular biologists, it is *the* creature for investigating evolutionary genetics. This insect is used because it is easy to culture in the laboratory, has a short generation time and is genetically relatively simple, having only four pair of easily observed chromosomes containing only 13,000 genes. If a new species were going to suddenly arise to support macroevolution there could be no better parent than the lowly fruit fly. Yet, sadly for evolutionists, quite the opposite has occurred.

Fruit fly, ***Drosophila melanogaster.***

In March of 2000, the big announcement was made that entire genome of the fruit fly had been sequenced. Radiation, such as x-rays, will produce mutations and various frequencies and strengths of x-rays have bombarded these insects in the laboratory, producing, for example, wing abnormalities known as *apterous*, *vestigial*, *dumpy*, etc. Since 1910, geneticists have documented over 3,000 mutations in this creature, yet science journals have not documented a single fruit fly evolving into something else, no matter how often they've mutated. The late evolutionist Pierre-P. Grassé said, ***The fruit fly (<u>Drosophila melanogaster</u>), the favorite pet insect of the geneticists, whose geographical, biotopical, urban, and rural genotypes are now known inside out, seems not to have changed since the remotest times.*** (Grassé, 1977)

As an embryo develops, its body plan arises under the direction of developmental control genes which includes a newly discovered group called the homeobox, or *Hox* genes. The *bithorax* gene is part of the *Hox* genes which, if mutated, may produce a four-winged fruit fly (they normally have two). It is said that "in many cases, experimentally induced mutations in homeotic genes create drastic changes in the [basic body design]" (Campbell, 1999) and one non-creationist stated, ***Control genes like homeotic genes may be the target of mutations that would conceivably change phenotypes, but one must remember that, the more central one makes changes in a complex system, the more severe the peripheral consequences become. . . . Homeotic changes induced in Drosophila genes have led only to monstrosities, and most experimenters do not*** *expect to see a bee arise from their [fruit fly] constructs* (Schwabe, 19??). Decades ago, an example of a "good mutation" was given by a University biologist during a public debate with Frank Sherman. It involved the *bithorax* gene that produces an atypical four-winged fruit fly. Unfortunately, the evolutionist failed to tell the audience that the fruit fly's ability to fly was severely impaired. Natural selection would have eliminated the mutated creatures quickly. Another hoped for and much needed "proof" of evolution fails after thoughtful consideration of the facts.

Honey Bees, Another Nightmare For Evolutionists

Next to the fruit fly, the most popular insect for the scientists is the honey bee. Much has been written and filmed of this insect's incredible ability to make perfectly formed beeswax combs containing hexagonal cells for maximum strength and storage space for honey, yet using the least possible amount of wax. This evidence of design plagued Darwin throughout his life. The bee's ability to convey the precise location of a new food source to fellow workers by a sophisticated "dance" is legendary. The vertical angle of their dance corresponds to the direction of the flowers to the sun and the rate of the dance depicts distance. The worker gives out samples and while she is still in the hive dancing, workers fly directly to the new food. This is an example of bees communicating abstract information and remains an enigma for evolutionists for they thought such skill developed only recently with primates.

Albert Einstein is attributed to having said, *"If honey bees become extinct, human society will follow in four years."* Many fear we are on the verge of this happening. Starting in the fall of 2006, bee colonies started disappearing in massive numbers and the strange disorder is called Colony Collapse Disorder or CCD. Already over 90 percent of the wild bee population in the United States have died out. The same thing is happening in Europe and India. Bee pollinated wildflowers in some regions have already started to decline by as much as 70 percent. Commercial beekeepers have now lost nearly 90 percent of bee colonies in twenty-two states. The bees simply leave their beehives and die. It is like a widespread mass suicide. Cell phone usage is implicated, but no one is certain as to the cause, nor is any remedy in sight. The situation is dire.

I've been keeping bees for over thirty years and can attest to recent problems in western Oklahoma. Five years ago, the so-called Africanized or "killer bees" arrived in Oklahoma and infected my apiary. In 2007, I lost three colonies of bees to CCD. In 2008, I lost another five colonies. This problem is real. Beekeepers are developing resistant strains, but they are not yet commercially available. Few people know this, but beekeepers purchase specific genetic strains of mated queen bees much as we do cattle or fruit trees. I always select the most aggressive bee varieties as they seem to produce the most honey and they add an element of excitement to my honey collecting.

The Disappearing Honey Bee.

Twenty-first century research has now revealed that bee vision is more complex than anyone had thought. According to science, arthropods have always been complex and their origin remains unknown. The first arthropod found in the fossil record is the amazing trilobite, common in Cambrian and Ordovician sediments. Many of these creatures are so well preserved that a detailed analysis of their eyes has been possible:

The elegant physical design of trilobite eyes employ Fermat's principle, Abbe's sine law, Snell's laws of refraction, and compensates for the optics of birefringent crystals. Thus, trilobites could see an undistorted image under water. Imagine being able to see with undistorted vision in all directions, being able to determine distance in part of that range, while, at the same time, having the optimum sensor for motion detection. (Austin, 1994)

So, from the beginning, arthropod vision has been extremely complicated, a feat impossible by undirected Darwinian evolution. Indeed, even explaining how the arthropod *head* supposedly evolved is an "acrimonious field." The composition of the arthropod head is one of the most bitter and longest-running problems in animal evolution. Unresolved after more than a century of debate, this tale is famously known as the "endless dispute" (Budd, 2005).

Typical Beehive

The brain of the bee is composed of a mere one million neurons (nerve cells), 0.01 percent of the neurons of a three-pound human brain. Using this tiny brain and associated vision, bees have been able to solve complicated color puzzles (*Astrobiology Magazine* Nov. 6, 2005) and even recognize human faces (Unger, 2005). They do this by using their 6,300 ommatidia that comprise the eye. Bees have also been created with the ability to distinguish up to three hundred separate flashes of light *per second*, an attribute they use as they rapidly fly over the changing landscape. The next time a bee buzzes by you on its way to a field of wildflowers, remember that it is designed to do and find things that our most sophisticated machines and computers cannot do, using vision and a brain that flies in the face (so to speak) of undirected evolution. See Smith (2007) for a detailed description of an unusual form of pollination that clearly displays evidence of design.

Fireflies, A Nighttime Delight

On a warm summer evening, fireflies entertain us with their amazing bioluminescence light show. Human delight is a byproduct of their searching for a mate. Each species has a specific pattern of flashes that identify them to the flightless female on the ground. This enchanting display is a complex form of light-without-heat called bioluminescence. In tropical waters at night, the wakes of ships also provide spectacular light shows from billions of single-celled luminescent dinoflagelates. The Brazilian railroad worm (actually the larvae of the large beetle *Phrixothrix*) sports a pair of red spots on its head and a pair of green lights along the sides of the body. Glowing mushrooms can be photographed by the light they emit. I recall a story of my own grandfather stopping his horse and buggy to walk into the nearby woods one night. A fallen tree was glowing so brightly from luminescent fungi, he thought it was on fire, but was amazed the glowing tree was cool to the touch. In some countries, lanterns using the light from fireflies are used. During World War II, Japanese troops crushed luminescent insects in order to read maps in the field. Biological luminescence is light produced by a complex chemical reaction within an organism. The chemical reaction producing this "cold" light involves two chemicals: "luciferin" which produces the light; and an enzyme called "luciferase" that catalyzes the reaction.

Charles Darwin was so enamored with bioluminescence that he admitted his theory could not explain its origin. Recent research has shown it to be commonplace in the oceans (Marchant, 2000). A Harvard biologist estimates that 75 percent of deep-dwelling animal species produce biochemical light. (Trombly, 1995) Many marine creatures use this essential light to find a mate, defend themselves against predators, or help them find food. Some shrimp were created with color filters, reflectors, and accessory lenses much like search lights made by man.

The weak unproven idea of "convergent evolution" is used in an attempt to explain how bioluminescence has evolved at least thirty different times. Such a widely held view has been proven impossible by probability and recent discoveries in molecular biology. The chance that two unrelated kinds of animals would have exactly the same result from random mutation is nil. It requires far less faith to view bioluminescence is evidence of an all knowing Designer. The Bible is clear on such matters. ***Through him all things were made; without him nothing was made that has been made.*** (John 1:3, NIV)

Bombardier Beetle Uses Chemical Warfare

This remarkable beetle has long been seen as evidence for creation and a continuing thorn in the side of evolutionists. The more we learn about this extraordinary insect and its uniquely complex defense strategy, the stronger is the evidence it was created. Evolution has failed miserably to explain the origin of its powerful and effective chemical warfare armament and accurate delivery system.

Bombardier Beetle

It involves far too many parts and complex toxic chemicals to have come about by random mutation. Such a complicated system shouts design at many levels and is indeed testimony to an all wise Creator-God. Let's consider a few of the remarkable facts.

There are over five hundred species of bombardier beetles known for their unusual and powerful chemical defense. They can accurately fire a boiling hot foul-smelling liquid at a potential enemy. The expulsion is accompanied by a loud popping sound and provides excellent protection. Bombardier beetles produce and store two powerful toxic chemicals, hydroquinone and hydrogen peroxide. When threatened, the two chemicals are forcefully squirted into a chamber where they are mixed with a catalytic enzyme. The chemicals undergo a violent exothermic chemical reaction resulting in the mixture instantly becoming hot enough to boil and are then sprayed on the threatening animal. The damage caused can be fatal to attacking insects and small creatures and is extremely painful to human skin. The reason such a complex and uncommon defensive mechanism is a problem for evolutionists is obvious. All of the components, including glands to produce, store and release the toxic chemicals along with the sophisticated and accurate method of mixing, aiming and expulsion must all occur together to be of any value. The development of any of the components' parts without the complete system would be of no value and would be eliminated by natural selection. Such a beautifully complex weapon shouts of the wisdom of its Creator while making a laughing stock of evolution dogma. Random genetic

172

errors could not invent such an incredibly complicated design, no matter how much time was available.

Cicada Killer, Surgeon Extraordinaire

Beautiful Yellow and Black Cicada Killer.

For the last example, let's consider an insect common to western Oklahoma where I grew up. It is one of largest North American wasps, the beautiful yellow and black cicada killer, ***Sphecius speciosus.*** As a boy, I watched many times as they captured and carefully sedated cicadas (called "locusts" in Oklahoma) which are about three times their size. First, they would systematically sting the nerves that control the flight muscles, rapidly subduing the huge insect. Next, one by one, they would sting the nerve ganglia associated with movement of each leg. Once the prey was effectively paralyzed, she would climb the nearest tree trunk or fence post dragging the huge cicada along behind. She would then fly off, clutching it tightly with their legs. The weight was too much and they were unable to maintain altitude and would gradually sink to the ground.

Upon landing, they would again drag the heavy prey up a nearby post or tree trunk and repeat the process until she finally returned to the previously dug burrow. She would then drag the prey underground and lay a single egg in it. The entire process would take an hour or longer. Years later, I found out the cicada must remain paralyzed, but alive to nourish the cicada larva. The larva instinctively eats around the vital organs until it is nearly mature. Finally, it eats the vital organs, killing the cicada and pupates. Even as a young boy, I was impressed at the instinctive knowledge the adult cicada wasp had of the cicada's nervous system as well as its ability to find the burial hole. The hand of God was clearly seen in His creation. I find it strange today with all we know about the complex structure and behavior that anyone can still cling to the belief that all this just happened without direction in small inheritable steps by random mistakes in the genetic code. Truly, as God's Word declares only the fool fails to see the hand of God in nature.

The fool says in his heart, "There is no God." They are corrupt, their deeds are vile; there is no one who does good. (Ps 14:1)

In summary, an examination of the scientific facts provides no evidence for the origin of animals by gradual change over time. Evolution is a view held by faith and is not supported by scientific facts. Instead, when viewed with an open mind, the evidence provides overwhelming support of the creation of animals in the present form by an all knowing Creator. There is also ample evidence supporting the global flood as the origin of the vast fossil record, but others have covered that subject far better than I could ever do. Readers are encouraged to do additional reading in these and related areas.

Chapter 5
Science in the Bible

He stretches out the north over empty space, and hangs the earth on nothing. (Job 26:7)

Earth From Space
NASA Goddard Space Flight Center Image by Reto Stöckli

The Bible is unique among all other books because it claims divine authorship. This gives the Bible authority and value other books do not have. Since it claims to be the infallible word of the living God, where it touches on science, history, or any other area, it must be without error. The unique truth of scripture has been verified repeatedly in each of the areas it addresses. From early childhood, I have been interested in the places science is mentioned in the Bible. Even as a child, I knew I was going to become a scientist and had confidence the Bible could be trusted. My interest in the relation between science and the Bible has deepened over the intervening years with education, maturity and a life spent doing and publishing scientific research. My confidence in biblical inerrancy has also increased with the passing of time. Perhaps nothing attests to the truth of the divine authorship of the Bible more clearly than the many places science is mentioned. Keep in mind the Bible was written centuries before science as we know it was established, yet the pronouncements have uncanny accuracy and relevance today.

Importance of Science

We live in a highly advanced technological society and science is important to all of us. Our lives are touched by science and modern technology in countless ways every day. Consider for a moment recent advances in medicine that enable organ transplants, pacemakers and even artificial hearts...all considered impossible a few decades ago. Genetic research has defined the entire human genome providing detailed understanding and treatment for a host of previously incurable genetic disorders. The Internet and cell phones provide instant personal communication with people all over the world. Global position satellite tracking (GPS) has revolutionized travel and even hiking in remote wilderness areas without fear of becoming lost. Exploration of space has exploded with man walking on the moon and science probes gathering detailed scientific information from Venus, Mars, and Jupiter as well as beyond our solar system. The Hubble telescope located above the earth's atmosphere has provided breathtaking pictures and a treasure trove of scientific information about our own galaxy and beyond for nearly two decades. The International Space Station, traveling 17,000 miles per hour and located 200 miles above the earth has provided a unique environment for scientists to live and perform research under zero gravity conditions for over ten years. Scientists from all over the world have been involved in this massive and unusually productive project. Additional workers, supplies and scientific equipment are provided by spacecraft from United States and Russia. Such national cooperation was unfathomable a short time ago during the cold war.

The relation of God's word to science has never been more important and relevant than it is today. It is important to know if the statements in the Bible dealing with science are trustworthy. Many today would emphatically say, "No." They would argue it is a moot point for people living when the Bible was written knew nothing of modern science. Others argue the Bible is not intended to be a science book, so its truth is such matters is unimportant. Yet God's Word claims to be true in all areas and not just regarding spiritual matters. If we can trust what it says regarding our salvation for all eternity, is it not logical that we can also trust it regarding history and scientific issues?

Those who argue from a historical perspective of what was known or not known at the time the Bible was written miss the point. God Himself inspired the Bible. It did not come solely from the mind of man. A few biblical passages will be repeated in order to emphasize this important point. *All scripture is inspired by God and profitable for teaching, for reproof, for correction, for training in righteousness: that the man of God may be adequate, equipped for every good work.* (2 Tim 3:16-17) There are more. *But know this first of all, that no prophecy of Scripture is a matter of one's own interpretation, for no prophecy was ever made by an act of human will, but men moved by the Holy Spirit spoke from God* (2 Pet 1:20-21). In addition, from Proverbs, *Every word of God is tested: He is a shield to those who take refuge in Him* (Proverbs 30:5). Where there is conflict between natural man's interpretation of science and the Word of God the choice is simple for the believer: *It is better to trust in the Lord than to put confidence in man* (Psalms 118:8). This passage is also the center verse in the Bible. There are as many verses before it as there are following. In a very real sense, one's acceptance of all of God's Word pivots on this important verse.

In many places, the Bible touches on science as it does history. There can be no contradiction between true science or true history and the Bible, because God is Author of both. He alone was there in the beginning of creation. He alone saw what happened and provided us with the only reliable record. Man and secular science can only speculate about how the world and life began. God's Word alone provides details and an eyewitness account. If one wants to learn how the universe, earth and all living things came into existence one must study the Genesis account of creation and the hundreds of other references to God as Creator in scripture. Let's look at what the Bible says regarding a few important areas of science.

Astronomy

Consider the following from the oldest book in the Bible, the book of Job. It was written before higher mathematics, telescopes, or the birth of astronomy or any branch of science. *He stretches out the north over empty space, and hangs the earth on nothing* (Job 26:7). The God of Creation inspired the writer who knew nothing of science or astronomy to make this remarkable statement. Keep in mind that what Job was saying was contrary to what learned men centuries later thought for they were confident the Earth was balanced on the back of a giant tortoise. This reptile was in turn walking (or swimming) slowly around the sun. Job wrote what God inspired him to write without question of fear of what men might say. There is no other rational explanation. It took secular science over 3,500 years and Sir Isaac Newton to describe this unique force called gravity that holds our earth the perfect distance from the sun. Indeed, it was God who hung the earth on nothing. How could Job have possibly known?

Consider another insightful bit of knowledge about astronomy that has only recently been confirmed by modern science. There is one glory of the sun, another glory of the moon, and another glory of the stars; for one star differs from another star in glory (I Corinthians 15:41). There is more. Consider the words of the Psalmist. He determines the number of the stars and calls them each by name. Great is our Lord and mighty in power; his understanding has no limit (Ps 147:4-5, NIV). To the unaided eye or even with powerful telescopes, many stars look the same. Today astronomers estimate there are on the order of 10^{21} stars in our universe. If you write that number out, it looks like this: 1,000,000,000,000,000,000,000. There a lot of stars, yet our God knows each one by name!

An amazing confirmation of what the Apostle Paul was inspired to write nearly two thousand years ago has occurred within the last two decades. This has been recently documented through detailed spectral analysis of the light of individual stars. Each star differs in color, brightness and temperature. Their spectral analysis and composition varies slightly from one star to another. Each star is unique as written long ago in scripture. What other religious writing would dare to say that each star in the universe is unique? Scripture plainly teaches the Living God called each star into existence by His spoken word. Only the God of Creation knows every star by name. God alone knew each was unique. Only God had such detailed astronomical wisdom for it was he that created them. Only the Creator could inspire men to write of the stars with such profound insight. This is truly convincing evidence of the supernatural authorship of the Bible. This is the same loving God that knows even the number of hairs on our head! (Matt 10:30, NIV). Hang on for there is more.

Consider the profound astronomical insight in the following passage of scripture: He sits enthroned above the circle of the earth, and its people are like grasshoppers. He stretches out the heavens like a canopy, and spreads them out like a tent to live in (Isaiah 40:22, NIV). Isaiah was probably not appreciated by the secular scientific community of his day. Here, the prophet is describing a spherical shape to our planet. Indeed, the Hebrew word is 'khug' and it refers to a compass. A study of this verse would not lead one to conclude Isaiah is describing a flat earth as was accepted by ALL learned people of that day! It would be three centuries later that Aristotle (On the Heavens) would suggest earth was spherical. Many more centuries would pass before this truth was widely accepted. The above scriptural passage clearly demonstrates two things. First, the Bible can be trusted where it mentions science. Secondly, such passages clearly show the Bible must have been inspired by the Creator Himself, for man could not have known these truths at that time without divine revelation.

Atomic Physics

Even basic atomic physics was alluded to in Scripture long before modern science postulated the modern view. Consider this written long before molecules and atoms were considered to be part of the mix: *By faith we understand that the worlds were prepared by the word of God, so that what is seen was not made out of things which are visible* (Hebrews 11:3). There is more. *He is before all things, and in him all things hold together.* (Col 1:17, NIV) Science has long known that opposite charges attract and like charges repel. It was relatively recently that physicists obtained experiential conformation of the complexity of the forces and particles that make up the nucleus of an atom. Once again these statements were once in doubt, but God inspired those accurate descriptions of atomic structure long ago.

Meteorology

Consider this accurate description of the hydraulic cycle, written nearly five thousand years ago: *For He draws up the drops of water. They distill rain from the mist, which the clouds pour down; they drip upon man abundantly* (Job 36:27-28). As this passage clearly demonstrates, what the ancients knew of science or the hydraulic cycle is irrelevant. God-the-Creator designed the hydraulic cycle and inspired the Holy book. Under supernatural influence, the author of Job recorded this amazing scientific insight long before it was understood by secular science. The science of meteorology was not refined enough in those days to understand the concept of water vapor or the formation of rain from vapor. It was unknown how water, once it had fallen to earth, could overcome gravity and return once again to the clouds. The Bible also mentions global atmospheric circulation, *The wind goes toward the south, and turns around to the north; The wind whirls about continually, and comes again on its circuit* (Ecclesiastes 1:6). Once again, God's word contains absolute truth regarding the science of meteorology. There can be no error, for it was God-breathed and did not depend on man's limited wisdom.

Biology

The Bible has more to say about biology than all other areas of science combined. No doubt, this is because God knew this portion of science would come under attack in these last days by evolution dogma. Darwin's unsupported idea of "descent with modification" is known today as macroevolution, and is at odds with the Genesis account of origins and with scientific observation. Scripture and modern science clearly describes animals, plants and people producing after their kind. This suggests there are natural limits to biological change. A cursory reading of Genesis shows like begets like, *Then God said, "Let the earth bring forth grass, the herb that yields seed, and the fruit tree that yields fruit according to its kind, whose seed is in itself, on the earth"; and it was so. And the earth brought forth grass, the herb that yields seed according to its kind, and the tree that yields fruit, whose seed is in itself according to its kind. And God saw that it was good.* (Genesis 1:11 and 12) *So God created*

179

great sea creatures and every living thing that moves, with which the waters abounded, according to their kind, and every winged bird according to its kind. And God saw that it was good. (Genesis 1:21) *And God made the beast of the earth according to its kind, cattle according to its kind, and everything that creeps on the earth according to its kind. And God saw that it was good.* (Genesis 1:25)

This is confirmed by modern genetics. We now understand how genes influence the expression of inherited traits. We also understand the genetic code to be a highly sophisticated and abstract written language. God not only created all living things in the beginning, but commanded them to reproduce "after their kind" or to follow their own unique genetic code. God can be seen not only as Creator but as Author of every living thing for it was He that wrote out their complex genetic code.

We also understand the number of genes associated with any given trait is limited beyond which farther genetic change is not possible. For example, milk production has been greatly enhanced by selective breeding, but a point is reached where milk production can not be farther increased and of course cows remain cows. They do not change into something different. Kinds do indeed reproduce after their kinds as stated in scripture. Let's consider another example, often forgotten.

For decades, entomologists thought migratory locusts had a compass sense like that of birds and whales. Locusts were thought to navigate using the earth's magnetic field and the position of the sun along with their own built-in biological clocks as they migrated. For a time, those scientists scoffed at God's word which clearly said the wind brought the locust infestation to Egypt long ago. As knowledge replaced ignorance, we now know migratory locusts have but one instinct...to fly straight up. They are totally at the mercy of the prevailing winds. Notice what God's word accurately stated this "new" scientific discovery over four thousand years ago. *So Moses stretched out his staff over the land of Egypt, and the Lord directed an east wind on the land all that day and all that night: and when it was morning, the east wind brought the locusts* (Exodus 10:12-14). God is Author of true science and of His Word. He created the locusts and the winds. Who better to know how they migrate? There can be no contradiction. Had Bible scholars sided with science decades ago, they would have been wrong, for the science of the day was in error. God's Word is true and *can* be trusted, even when it speaks of science.

Human Health

The Bible also has a great deal to say about the human body and especially health issues. Regarding the amazing circulatory system of man and beast, creation scientist William Harvey, in 1616, verified what Scripture had clearly mentioned three thousand years earlier. *For the life of the flesh is in the blood, and I have given it to you upon the altar to make atonement for your souls; for it is the blood that makes atonement for the soul.* (Leviticus 17:11) Life is indeed in the blood for it alone carries life giving oxygen throughout the body and removes toxic wastes. Without blood, all tissue quickly dies.

It is well established that our mental state can drastically alter the way our body functions. Worry can literally kill us. According to some estimates, psychophysiologic illness, formerly called psychosomatic illness, accounts for as much as 80 percent of visits to one's physician. Although they often have their origins in the mind, they are real and treatment is often required. They should not be confused with hypochondria, which is a mental disorder where a person has a preoccupying fear of having a serious illness. The conviction persists despite appropriate medical evaluation that the person is in excellent health. Psychophysiologic illness can include asthma, allergies, false pregnancy, celiac disease, peptic ulcers, debilitating depression, migraine headaches and a host of other maladies. These systemic effects occur because of the close association between brain and body. That pathway is the hypothalamic-hypophysial portal system that carries releasing factors from the lower part of the brain, the hypothalamus directly to the pituitary or hypophysis gland. In response to these minute quantities of releasing factors, the pituitary gland releases its hormones that influence all parts of the body. It is for this reason that our mind influences how our body functions. The important role of the mind over the human body is clearly described in Proverbs. *A cheerful heart is good medicine, but a crushed spirit dries up the bones* (Proverbs 17:22). It is well known today that laughter or even petting a dog or cat can lower blood pressure, increase the immune response and bring a host of other important benefits. Laughter therapy is growing in popularity around the world. In contrast, worry and depression can lead to illness and death. Again, the role of pleasing words to good health was mentioned long ago in scripture. *Pleasant words are a honeycomb, sweet to the soul and healing to the bones.* (Proverbs 16:24, NIV) This profoundly important connection has only been recently confirmed by modern medicine, but has long been on the mind of God and in is in scripture.

Healthy Food

The wisdom of God and His concern for human health can be also seen in many of the laws in Leviticus dealing with what could be eaten and what was forbidden to eat. For example, swine were considered unclean. We know today that such animals are omnivores and will eat almost anything. As a result, they often have a very high rate of parasite infestation. Without proper precautions during butchering and cooking, eating such animals would greatly increase the risk of parasite infestation and illness in humans. Many of the laws related to cleanliness are clearly seen as important in avoiding contact with animals that had died of disease or other causes. Here is one of many examples that could be given. *Or if a person touches anything ceremonially unclean-- whether the carcasses of unclean wild animals or of unclean livestock or of unclean creatures that move along the ground-- even though he is unaware of it, he has become unclean and is guilty* (Lev 5:2, NIV). Obviously, any animal found dead would be a threat to human health and should be avoided.

Conclusion

More examples could be given from the science in the Bible, but the point is obvious. God's Word is trustworthy when it touches on science because God is the Author of both His inerrant word and true science. The examples of science revealed in the Bible are true, not because of what the men who wrote the words knew at the time, but because of God who inspired the words. To conclude this section, whether one probes into the deepest reaches of space, or examines the functions of the smallest living cell God's handiwork is clearly seen.

As Christians, we must interpret the facts of science into a scriptural framework. To allow Scripture to follow secular science leads to the dangers of theistic evolution (See chapter 9). Non-believers need to understand the Bible is true even when it touches on science or history. Without acceptance of the Bible as true everything changes. For example, even the Roman Road to Salvation through Christ becomes meaningless if there was no perfect creation and subsequent fall into sin. We *must* know not only what we believe, but also why we believe it. We must be willing to defend it and share these truths with others. God's Word is true and can be trusted in all areas. Science is the search for truth and truth leads directly to the God if creation. Jesus Himself claimed to be the Truth. *Jesus answered, "I am the way and the truth and the life. No one comes to the Father except through me"* (John 14:6, NIV). With truth comes freedom. *Then you will know the truth, and the truth will set you free."* (John 8:32)

Chapter 6
Secular Science and the Bible At War

All things were made through Him, and without Him nothing was made that was made. (John 1:3, NKJ) Ears that hear and eyes that see-- the LORD has made them both. (Prov 20:12)

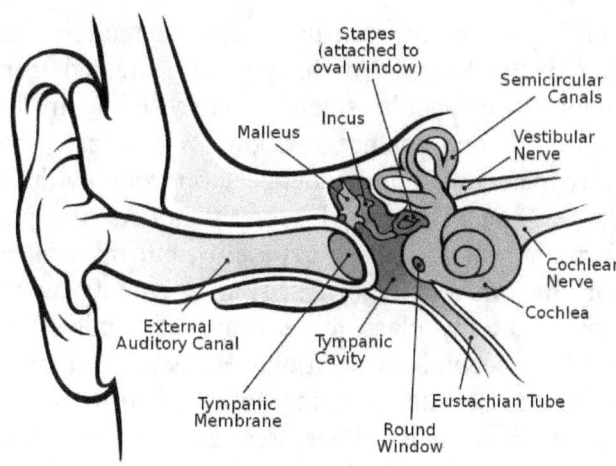

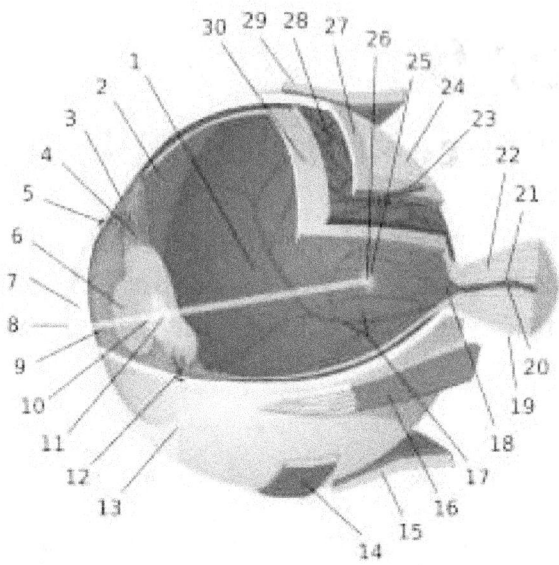

Human Ear and Eye from Wikipedia

Anyone failing to see the conflicts between modern secular science and the Bible is either badly deceived or has recently arrived from another planet. The contradictions between secular science and the Bible are blatantly presented in university classrooms and science textbooks. I actually had a professor in graduate school say there was more evidence supporting evolution than there was supporting gravity. I must confess I secretly hoped that particular professor would test that hypothesis by jumping off a cliff, but my wishes went unheeded. Another professor, this one at a major Christian university, would ask on the first day of comparative anatomy class for a show of hands of any students who believed in the biblical account of creation. He would remember those students and openly ridicule and belittle them the entire semester for their ignorance in believing such an outdated myth. Peer pressure and the willingness to become immersed in their chosen area of science often compel students to conform to their professor's beliefs even if they differ from God's Word or their own Christian upbringing and values. They are in the university to learn and often feel they must learn and accept the views of their learned professors. One is reminded of scripture, ***Although they claimed to be wise, they became fools.*** (Rom 1:22).

After spending much of my adult life associated with secular universities, I know many Christian students begin doubting their faith when exposed to godless evolution because they were not properly grounded in the Word of God. Nor do most of our churches teach there can be *no* conflict between true science and the Word of God, because God is the Author of both. As a result, students are not prepared to think through the persuasive arguments by their professors or what they read in textbooks. As classroom attacks continue, many begin doubting larger portions of God's Word. They no longer view the Bible as inerrant and some end up believing Christianity is no different from any other world religion. Others take the ostrich approach and are afraid to face the issue. They flee all science courses and change their major to avoid confrontation. I have seen this happen far too many times and no amount of persuasion could change their minds. This is not a solution. Their doubts linger and haunt them for the rest of their lives, weakening their faith in the Bible. Jesus did not criticize Thomas for doubting, but helped him resolve his doubts and he traveled farther than any of the other apostles and started a church in India, The Church of the Martyr that has survived until today and sends missionaries to the United States. We must be willing always to help students resolve their doubts about the truth of scripture.

But, Isn't Science Important?

As mentioned previously, we live in a scientifically and technically advanced society. Our lives are touched by science and technology everyday. Science has given us many good things. However, we must always put this into proper perspective and remember God gave man dominion over the world and science is a part of man's dominion. Yet to many, science is more for it has become a god. To doubt any aspect of science in the current culture is akin to sacrilege. In the 1800's, preachers were often the most educated individuals in the community and were held in high esteem. Today, scientists and university professors have the most education and influence, while preachers are often ridiculed as unlearned, superstitious or worse. It is becoming increasingly common for preachers in television sitcoms and movies to be portrayed as religious nuts and villains. Certainly, the highly publicized fall of several widely known television evangelists and other prominent Christians has worsened that image. Today, the Ten Commandments, prayer and the Bible are systematically being removed from schools and public places. The mention of the name of Jesus is forbidden in our public schools, yet students are taught Muslim prayers and must learn the five pillars of Islam. Christians are considered "dangerous intolerant fundamentalists" and are equated by some with the Taliban and other terrorist groups. In a very real sense, many scientists and university professors worship at the altar of science. Such scientists believe given enough research money, science can solve any problem and cure every illness. They need only visit a veteran's or children's hospital to see that modern medicine has not ended all suffering and cannot cure all physical maladies.

To make matters worse, our society is increasingly accepting relativism. Absolute truth and morals are rejected; even ridiculed. The temptation is for students to side with the majority godless worldview in science. Science majors and others are taught everything in the universe is natural and undirected. Life is nothing more than matter in motion and mankind is the result of mutational accidents. They are repeatedly told there is no scientific evidence for or need for the supernatural. Humanity, they are told, has advanced beyond a need for God. As a result of this endless droning, many begin to doubt Biblical truth and some will reject the Christian worldview for the rest of their lives. We must remember as Christians we are in the minority. The majority of truck drivers, store clerks and university professors have rejected God. *The fool says in his heart, "There is no God." They are corrupt, their deeds are vile; there is no one who does good. The LORD looks down from heaven on the sons of men to see if there are any who understand, any who seek God.* (Ps 14:1-2) We are to be salt and light and must not buckle under societal pressure.

Common Excuses

There are two commonly heard excuses for rejecting Biblical teachings touted by "intellectuals." These excuses are even taught at some or our leading seminaries! Neither excuse holds up under careful scrutiny. Perhaps the most commonly heard excuse is, "The Bible was not intended to be a science book." The other common reason given for not accepting biblical truth is, "The ancient people in Bible time knew nothing of modern science." Nonsense! Both of these excuses miss the point of supernatural revelation of scripture. If the Bible was supernaturally inspired as it claims to be, then there can be no error where it touches on science, history, morality or anything else. For God is the Author of all true science and His Word. What the ancient people knew of science is irrelevant. Unlike all other books, the Bible was Spirit breathed...*for prophecy never came by the will of man, but holy men of God spoke as they were moved by the Holy Spirit* (2 Pet 1:21, NKJ). The secular world cannot accept this important Christian truth.

A Closer Look

As Christians, we must look closely at the apparent conflicts between modern science and God's Word and consider a reasoned resolution. If we can trust the Bible regarding the more weighty matters of heaven, hell, sin and eternal judgment, then it must be without error on less important matters including history and science. If the Bible is the infallible Word of God, there can be only two explanations for conflicts between the Bible and secular science: error in Biblical interpretation or error in the interpretation of secular science. Sadly, many higher education institutions dismiss the Bible as nothing more than a collection of fables from the unlearned.

Two pictures emerge. Modern science is filled with the firmly held dogma of godless evolution and is presented in high school and college textbooks as established and proven scientific fact. High school and university teachers support this fable largely from ignorance. Few have examined the actual evidence closely. Such tightly held evolution dogma is contrary to the widely held view that science is objective. It is not. Scientists see the world though a lens tinted by their own beliefs as does everyone else. As we shall see in the next chapter, the actual evidence supporting macroevolution is virtually non-existent. Evolution is accepted by faith and provides the foundation for secular humanism which has replaced Christianity as the religion of our nation.

There are two more reasons such considerations are important. Many feel science has disproved Genesis. This is tragic, for much of our Christian faith rests on the first eleven chapters of Genesis. Perhaps more than any other part of the Bible, it shows God's love for his people and his enduring patience. It provides a history of God's chosen people and their rebellion. Without the fall into sin there is no need for a Savior. If Genesis is rejected, doubts are raised regarding the entire Word of God. Such doubts keep believers from spending time in the Word. We see a systemic growth of such doubts not only in individuals, but also in many Christian denominations today. First, aspects of the account of Creation are doubted, especially the Genesis timeframe. Next, some of the miracles are thought to be myth until finally they doubt the truth of Bible itself and even the existence of God or the need for a Savior. Some end up believing Christianity is no different from other world religions. The Body of Christ must wake up, for without a firm foundation, "What can the righteous do?"

Lingering doubts weaken and destroy. Resolved doubts strengthen and build. The Apostle Thomas doubted the bodily resurrection of Jesus Christ. Jesus did not criticize him for doubting, but helped him resolve his doubts (John 20:24-28). After his doubts were resolved, he went farther geographically than did any of the other apostles. He traveled to India and started the Church of the Martyr that remains influential to this day sending missionaries throughout the world including to the United States. My prayer is this discussion will help readers to resolve doubts regarding the truth of God's Word as it relates to science. The stakes could not be higher. For many dealing with this issue, eternity hangs in the balance.

The other reason relates directly to evangelism. Many today that see God-the-Creator as myth will not accept God-the-Savior. If science has disproved Genesis, why should they trust the gospel message? Again, I have seen this scenario repeated for countless university students. For lost science students, the teaching of godless evolution erects barriers blocking *all* roads to the Cross. Only intellectual arguments can remove these roadblocks and lead to Salvation. In our own personal spiritual growth and in witnessing to others, especially those interested in science, we must have confidence in the Word of God and be able to defend it intellectually. The approach I have heard some preacher's take is that, "It's in the Bible, and that is good enough for me" is totally ineffective in witnessing to university students and educated adults. Scripture tells us that we

187

are to: ***Always be prepared to give an answer to everyone who asks you to give the reason for the hope that you have. But do this with gentleness and respect.*** (1 Pet 3:15)   This means for some people, addressing some of the difficult intellectual arguments may be necessary in winning souls. Nothing is of more importance. ***The fruit of the righteous is a tree of life, and he who wins souls is wise. If the righteous receive their due on earth, how much more the ungodly and the sinner!*** (Prov 11:30-31)

Can the Bible Be Trusted?

I understand that to someone outside the faith, one cannot use the Bible to prove Biblical inerrancy. There are many excellent scholarly works on Biblical inerrancy and scientific apologetics. Let me encourage you to dig deeper. Such arguments are beyond the scope of this discussion. Christians understand God's Word to be true. Again, if we trust it for our salvation, can we not trust it when it touches on our origin, history and science? Where there is conflict between man's science and God's Word, Christians choose to accept the Word of God and discover where science has erred.

Warnings About Science

This is perhaps the most important, yet the most difficult portion of this chapter to accept. Some will not want to consider it. Let me make it very clear I am *not* anti-science. I have spent most of my professional life doing scientific research and have taught science to thousands of university students in classrooms and online. I have published technical papers and given talks about my research findings at scientific meetings throughout the United States, as well as in Canada, Brazil and England.

Science was part of man's God-given dominion over Creation, but science, like other blessings of God, has become perverted by man's sinful nature and by the Enemy.   God's word clearly warns us about the dangers of science in these last days. ***O Timothy guard what has been entrusted to you, avoiding worldly and empty chatter and opposing arguments of what is falsely called "science"- which some have professed and thus gone astray from the faith*** (1 Tim 6:20-21). Paul warns us in 2 Tim 3:7 that in these last days difficult times will come when men will be...***Always learning and never able to come to the knowledge of the truth.*** Could there be a clearer word picture of our society? We live in a time where knowledge is exploding, yet society has never been farther from the truths of God. In the nineteenth century, it took about fifty years to double the world's knowledge. Early in the twenty-first century, the world's knowledge is doubling in less than a year. Is it any wonder many call this the Information Age?

We are warned specifically in the Bible to beware of false philosophies: See to it that no one takes you captive through philosophy and empty deception, according to the tradition of men, according to the elementary principles of the world. (Colossians 2:8). Being warned is good, but there is more. We are to expose these deceptions and are to warn others. And do not participate in the unfruitful deeds of darkness; but instead even expose them. (Ephesians 5:11). Jesus was the light in a dark world. When Jesus spoke again to the people, he said, "I am the light of the world. Whoever follows me will never walk in darkness, but will have the light of life." (John 8:12, NIV) As Christians, Jesus has commanded us to be both salt and light in the world. You are the salt of the earth. But if the salt loses its saltiness, how can it be made salty again? It is no longer good for anything, except to be thrown out and trampled by men. "You are the light of the world. A city on a hill cannot be hidden. (Matt 5:13-14, NIV) Part of our responsibility as Christians is to be the salt and light in the world. That means, among other things, that we are to expose the lies of the Enemy. Today, godless evolution is one of the most pervasive lies from the enemy. Get the facts. Look closely at both sides of this issue. As Christians we have the Word of the Living God and scientific evidence on our side.

Evolution has also become a modern form of idolatry as we shall see in chapter ten. With the acceptance of evolution, nature and natural law are given credit for the origin of life as well as the diversity and complexity of living things. That credit and worship should go instead to the Creator of nature and natural law as clearly stated in Genesis and many other places in scripture. ***Through him all things were made; without him nothing was made that has been made.*** (John 1:3) In sharp contrast with scripture, evolutionists believe in the beginning was matter and energy and all living things came about without the need for a Creator. One of my motivations for ministry in the area of science and the Bible is to warn others of the errors in modern science, especially regarding moral implications of godless evolution. We must be informed and inform others.

For those who see no moral implications in evolution, please wait until we hear from some of the world's evolutionists as to why they accepted evolution. Many admit it was not the preponderance of scientific fact (in spite of what is taught in college) that caused them to accept evolution, but instead was at the core a moral issue. Their lifestyle demanded the removal of God from their own consciousness. Evolution did exactly that. Darwin's 1859 book was popular not because it was scientific, but because it explained creation *without* a Creator. Many people hostile to Christianity in Victorian England and many today needed this excuse. Darwin's evolution held the key and that is why it was so quickly embraced. It was not due to its scientific merit in spite of what we have been told. Finally, as Christians we must know why we believe what we believe and be ready always to defend these beliefs. This is never easy. Far too many Christian young people simply attempt to put on their parent's morals and beliefs without understanding the basis for such beliefs. Such individuals are not prepared for the attacks that will occur when they attend the university. Keep in mind we are instructed in scripture *to be prepared to give an answer to everyone who asks you to give the reason for the hope that you have* (2 Pet 3:15). As Christians we all have the hope, but many are not motivated or properly trained to give the reason for that hope. We must be engaged intellectually and always be ready and able to share the reason for our hope. This means intellectual debate. The stakes could not be higher, for many souls are at risk for lack of an intellectual reason to believe. Let us begin with renewed vigor and dedication to learn and share the reason for our hope and understand the lack of evidence for evolution. Truly, our Christian faith begins with those powerful words, *In the beginning God created the heavens and the earth.* (Gen 1:1) Without the God of creation, there would be nothing for *all things were made by him; and without him was not anything made that was made* (John 1:3, KJV).

Chapter 7
Scientific Evidence for Evolution

Have nothing to do with godless myths and old wives' tales; rather, train yourself to be godly. (1 Tim 4:7)

Abraham Lincoln and Charles Darwin were born on the same day, February 12, 1809. Abraham Lincoln is remembered for freeing the slaves. Charles Darwin should be remembered for enslaving the free. It is beyond comprehension how one of the best educated cultures in history can continue to allow the teaching of false information about evolution in public schools and university classrooms. Secular humanism, with its foundation rooted deeply in the sterile soil of evolution, has become an increasingly pervasive religion in America, replacing Christianity. Evolution continues to be accepted by faith for it is certainly not grounded in solid scientific evidence. The acceptance of evolution has always been a moral issue and remains so today. Evolution removes God from the world and from human conscience. Evolutionists define evolution as a fully natural process, leaving no room for God or any form of intelligent design.
The biblical kinds were commanded to reproduce "after their kind" (Genesis 1:11, 12, 21, 24 and 25) and the scientific evidence today supports this for all animals and plants. Solid evidence for any major taxonomic "kind" of plant or animal ever becoming another "kind" is lacking, yet such an admission is not seen in textbooks or heard in biology lectures. The only solid evidence for "evolutionary changes" is for small genetic changes such as antibiotic resistant bacteria, pesticide resistant insects, color change in peppered moths and the well-documented changes in domestic animals and food crops due to selective breeding over the centuries. Such changes are within the originally created kinds and offer no evidence in support amoeba-to-man macroevolution. Nor do they provide any information or evidence regarding the origin of life. There are many excellent scientific apologetics available and a detailed discussion of this topic is beyond the scope of this book. A few points will illustrate the weakness of the scientific evidence supporting macroevolution.

Even the casual observer can see the weaknesses of the standard textbook evidences put forward to support evolution. It is unsettling that blatantly fraudulent drawings and arguments continue to appear in high school and university biology textbooks decade after decade. It makes me angry that science authors and publishers continue to lie to young people about evolution. It is even worse that our learned professors continue to disseminate misinformation. They should know better. It is yet another evidence that evolutionists lack any actual evidence for macroevolution. It is for this reason they continue to use weak and even blatantly dishonest and erroneous arguments. Only a perverted sort of fervor to conceal the facts can account for the continued dissemination of misinformation in our public schools and universities regarding the origin and diversity of living things. They must be challenged to present arguments for and against evolution. The must openly discuss the actual scientific evidence even if it leads to the admission of design.

Discussion Is Forbidden

In spite of what science students are taught, modern secular science is not objective. Scientists are not free to follow the evidence if it leads to design or to the Creator. Instead, students are forced to accept evolution dogma based on untrue information. Dissent, counter arguments or evidence against evolution are strictly forbidden. Any discussion against evolution, even by well trained and respected scientists is not allowed. The discussion is closed. Why would any author use indefensibly false arguments or outright lies to support evolution unless that false information offers the best argument available? Secondly, and this is at the heart of the issue, one after another leading evolutionist, when pressed as to why they so eagerly embraced evolution, fail to mention the scientific evidence. Instead, they say it was a moral issue. Consider this point carefully for you will not see it in biology textbooks or hear it in science classes. Our students today are being lied to without apology. Why are evolutionists afraid of the actual scientific evidence? As we shall see, the answer is obvious for the actual scientific evidence fails miserably to support macroevolution. Science as we know it will die if scientists are not given freedom to follow the evidence no matter where it leads. Objectivity must return to science for it is the very cornerstone of discovery.

For our discussion, we need a working definition of this process called evolution or, more precisely, macroevolution. Some textbooks define evolution as simply "change." This is a cop out. Of course, living things have the ability to change or adapt to changing environments, but such a broad term lacks meaning is so vague as to be unless. All evolution is change, but not all change is evolution. Let us define macroevolution more broadly. I accept the following definition used by many leading evolutionists. Evolution is the natural process by which life arose from non-living material and the process by which all living and extinct species of plants and animals, including man arose. Evolution is generally defined as a change toward increased complexity (amoeba to man). This is macroevolution as opposed to microevolution. One is not the logical outgrowth of the other.

According to evolution, there are thought to be a few changes to less complexity (lizards to snakes and the loss of flight in some birds and loss of eyesight in certain cave dwelling species of fish), but these are considered exceptions to the rule. Remember, evolution is always defined as a fully natural process. If there was any supernatural intervention or direction, the process is not evolution. This point cannot be over emphasized for it clearly means the theistic evolutionist and those supporting Intelligent Design lack support of the established secular scientific community.

Kinds Remain Fixed

Some Bible scholars (including James Strong) have erred in trying to equate the Biblical "kind" (Miyn or Meen) to the scientific "species." This is unwarranted for several reasons. Few taxonomists today can agree on a precise definition for a species that fits all living things. There are also fundamental differences in classifying plants and animals. One size does not fit all. Since taxonomists cannot agree on the precise meaning of the term, it is useless for Bible scholars to say it means "species" in scripture.

It is also unwarranted from Biblical passages such as Lev 11:16-17 where the children of Israel were instructed not to eat three kinds of owls. According to modern taxonomy, there are over two hundred species of owls. It is obvious the Biblical kind is more inclusive than the biologist's ever-changing definition of species. Perhaps the Biblical kind is closer to what taxonomists call a genera or some other larger grouping of similar species. A good working definition for an animal species is a group of actually or potentially interbreeding animals, but again the rules change from one major group of animals to another and the definition has changed over the decades. Plants and organisms that reproduce asexually require a more complex definition. The recent cloning of livestock and other animals also clouds the issue. Again, the serious reader is encouraged to dig deeper for there is a vast scientific literature on this important topic both from creation and evolution viewpoints. Genetic information has literally exploded the past few years with the human genome project and especially recent discoveries of the details of how genetic information is stored and translated into structure and function. Much more will be learned during the next several years. Genetics is an exciting portion of contemporary science.

Two Kinds of Evidence for Evolution

If an attorney were to look at the seemingly endless variety of evidence used in textbooks and lectures to support evolution, he or she would divide the evidence into two broad categories: direct and indirect. Most evidence seen in textbooks and used in lectures to support evolution is indirect or circumstantial. As in a criminal investigation, where the evidence is circumstantial, alternative interpretations are possible. Such evidence is often ambiguous which is why direct evidence is preferred to prove guilt or innocence of a person in a court of law. It is the same with the evidence used to support evolution. Alternative interpretations of the circumstantial evidence are possible and will be discussed below.

As Christians, we must interpret the facts of science into the unchanging scriptural framework. Far too often, Christians have taken the opposite approach of attempting to interpret Biblical teachings into the ever-changing framework of modern science. If the evidence is real, we must deal with it. If the evidence is false, it should not continue to appear in high school and university textbooks. At least such "evidence" should be presented as speculation and subject to alternative interpretations. Some might raise the obvious question, "But aren't scientists objective?" This is an important point and understanding the answer is crucial to any discussion of origins. The answer is an ambiguous, "Yes and no."
If scientific experiments are well designed, the results should be clear. Statistical analysis is used in an attempt for objective interpretation. Regarding origins, we must always depend on indirect evidence and subjective interpretations. Man was not there when the original life appeared and the process cannot be duplicated in the laboratory. Attempts have been made for decades to do this, but thus far none have succeeded. Such attempts may be successful in the future, but this in no way proves it happened under natural conditions. Man can create an aircraft to fly safely, but only a fool would expect to find one in the aftermath of a tornado. Such events are destructive, not creative. Again, such experiments and their interruption are based on a multitude of assumptions and are beset with insurmountable problems. One thing is clear, the more details we learn about life processes the more complex it is. The same is true for the complex structures inside each living cell. These two facts are raising serious doubts that the origin of life could occur, no matter how much time was involved and has given rise to the Intelligent Design field. Scientists are human and bring their own prejudices and preconceived notions to the laboratory. What each of us see is colored by the lens of our education, experience, and especially by our own personal beliefs and by our worldview. So it is with scientists. We must examine the evidence carefully and draw our own conclusions. The stakes could not be higher. Our worldview regarding origins influences our lives and those around us now and for all eternity.

National surveys repeatedly indicate approximately 80 percent of parents want creation to be taught in our public schools, yet such teaching is now banned by law. The Ten Commandments and prayer has been banned from our public schools. Evolution is touted as the only scientifically defensible explanation for the origin and complexity of life. Students are taught man is the product of mindless evolution and life after death is a myth propagated by the superstitious and the unlearned. In spite of such brainwashing, a recent Zogby poll on the eve of Darwin's two hundredth birthday in 2009 showed a sharp increase in the number of Americans who wanted both sides of the evolution controversy discussed in biology classes. Over 78 percent wanted the academic freedom to discuss the strengths and weakness of evolution. This is up significantly from 69 percent in 2006. Evolution is rapidly becoming a failed theory both among scientists and the general public. Evolution like the passenger pigeon is destined for extinction.

In spite of the blind allegiance to the mantra of the so-called "educated elite," evolution has a surprisingly large faith element. Unfortunately, many science teachers tend to lump all theories together. For example, cell theory, atomic theory, relativity and evolution are often treated as equally valid and proven. This is simply not true. As previously mentioned, evolution is never taught as theory in high school or university biology classes. One of my own graduate school professors claimed there was more evidence supporting evolution than gravity. Nonsense! Evolution is accepted in large part by faith based on the worldview of some leading scientists past and present. Biologists and others want it to be true to evolution dogma by ignoring or even denying contradictory evidence. They simply will not or cannot accept there may be a Creator-God to Whom they are responsible. Again, we see the crux of the evolution issue is at the core a moral issue, yet this is never mentioned in the classroom.

Physics, chemistry, and in a sense mathematics make up the so called "hard" sciences and as such are falsifiable by well designed experiments or rigorous calculations. That is, laboratory experiments, having certain results, could be conducted that would falsify such "laws" and cause them to be abandoned. The laws of physics and chemistry have withstood millions of experiments and decades of testing and have survived the test of time. This is not the case with evolution.

Unlike the "hard" sciences and mathematics, evolution is not readily falsifiable. No single experiment can be designed that could clearly disprove evolution as is the case with the foundations for the other sciences. Instead, evolution is in large part a belief system accepted by faith that attempts to account for the origin of life and the diversity of all past and present species, by natural processes alone. It is an assumption and must be accepted by faith and not by empirical experiment or other direct scientific evidence. Since life is not currently arising from non-living material, nor are major kinds of plants or animals becoming something entirely different, evolution cannot be proven with the same certainty of the laws of physics and chemistry. It must therefore be based on indirect evidence and to a degree not seen anywhere else in science, it must be accepted largely by faith. Indeed, many hold to evolution with a religious-like fervor and zeal not seen in other areas of science. As is the case with religion, emotions are high and a rational discussion of origins is often difficult or impossible.

Yes, Christianity for some is accepted by faith. There is, however, a vast cosmological, anthropological, teleological, historical, archeological, traditional, as well as biblical evidence for the Christian worldview. Personal experience alone adds credibility far beyond laboratory experimentation. For example, many Christians know without a doubt that God is real because of what He has done in their lives as individuals. Certainly, faith plays an important role in believing in the God of Creation, but it is also based on solid scientific evidence.

Evolutionists, the news media and general public often overlook the fact there have been literally tens of thousands of published scientific studies for decades that lend credibility to the creation view of origins and to the global flood described in Genesis. The scientists who conduct these studies are well qualified and have terminal degrees from accredited universities. Many of them also write technical journal articles in other areas of expertise, as do I. Such scientific studies are every bit as testable by modern science as is evolution, yet the media or university professors seldom discuss this point. With this brief introduction in mind, let us look at the direct evidence used to support macroevolution. As you will see, this will not take long for the direct evidence is meager indeed.

Direct Evidence for Evolution

The amount of actual verifiable direct evidence for evolution is surprisingly small. It is this paucity of hard evidence that convinced me many years ago, that evolution was a belief system and not science. That view has deepened with education and time. For decades, most textbooks discuss the same old two or three direct "proofs" of evolution. There seems to be no new evidence and these "proofs" are extremely weak, as we shall see. The first is the well-known peppered moths of Great Britain and the second is the oft-evoked pesticide tolerance of parasites and insects. The other "proof" of evolution is the resistance of various pathogens to certain antibiotics.

Black and White Peppered Moths

The peppered moth is indigenous to England and has long existed in two common morphological types: dark and light colored. There are also intergrades, but these are seldom mentioned in textbooks. The evolution of peppered moths is no longer seen as an icon of evolution. A few years ago, this issue was debated ad nauseam, and the Darwinists lost badly. It is now widely accepted the whole argument was false and based on fraudulent "data." The moths shown in the infamous photos once seen in biology textbooks were actually glued to the tree. For this and other reasons, the peppered moth argument as evidence supporting evolution has disappeared from many new biology textbooks. Yet it remains a talking point for many university professors. This dead horse needs to be resurrected for completeness as many are still exposed to this fictional tale.

Following are typical photos of the two contrasting color phases often shown in biology and evolution textbooks. Again, this argument is weak indeed, yet many evolutionists and teachers still cling to it for lack of more substantial evidence evolution has occurred.

Light and Dark Phases of the Peppered Moth

Like other moths, peppered moths are active at night and often rest on tree branches or trunks during the day. During the early 1800's, or so the story goes, peppered moths rested on light-colored lichens that grow on tree trunks. Nearly all the moths collected during this period were light-colored. Only a few dark colored moths were collected as they were seen as easy targets for hungry birds. In the mid-1800's, however, factories burned so much coal that soot settled over the countryside, killing the lichens and blackening the tree trunks. Light-colored moths on dark-colored trees were easily seen and eaten by birds. As a result, more of the black moths survived and produced offspring. Within fifty years, most moths in heavily polluted areas were black. After anti-pollution laws were passed in the mid 1900's, the soot gradually disappeared and the tree trunks again became lighter as lichens returned and the number of light-colored moths increased. This is one of the standard proofs given in older textbooks that evolution has actually occurred. Many biology professors still accept to this outdated and proven erroneous argument.

To accept and perpetuate this as proof of macroevolution is naive...or worse; it is dishonest. Both the light and dark peppered moths are still peppered moths. No new specie has evolved! How mere color change can be touted as "proof positive" of evolution is astounding. Even if the above scenario were true, one can alternately see this as evidence that a wise Creator would foresee the need for color change with future pollution. Such limited genetic drift is often real, but can hardly explain the origin of a new species, new organs or the origin of life itself. God placed limits on genetic variability for he commanded that each newly created kind would reproduce only "after its kind" and it remains true today.

As described above, fraud regarding the alleged evolution of peppered moths was rampant in the original papers used to support evolution. What is shocking is those fraudulent photos appeared for decades in high school and university biology textbooks. Thankfully, they are no longer used in most current textbooks. Again, it is important to describe this evidence to illustrate the extreme weakness of the arguments used to support evolution and how very resistant many biologists to admitting the evidence was simply untrue. We will next examine the evidence of pesticide resistance for this, too, is often presented as a positive proof evolution is occurring.

Pesticide Resistance

I have lived most of my life on the family farm in western Oklahoma. I am well aware that flies and other pests develop resistance to various control methods. Flies and mosquitoes develop resistance to the chemicals used to control, repel or kill them. Farmers and ranchers must change to different kinds of fly repellent ear tags for cattle every few years. Yes, organisms were created with the ability to respond to environmental changes, but no new specie has magically arisen. The pests remain the same species of fly, worm or mosquito. There is absolutely no evidence of macroevolution in this common adaptive response. Such resistance is not evidence of how life arose in the first place or how millions of species of plants and animals could have evolved from that first living thing. Kinds still reproduce after their own kinds as commanded by their Creator during Creation week.

Antibiotic Resistance

Likewise, evidence abounds for antibiotic-resistant bacteria. Again, my professors taught this as proof positive of macroevolution. It is not. A few years ago, it was widely reported that a person had to have a lung removed because the strain of infecting tuberculosis bacterium had become unresponsive to every known antibiotic. No doubt, this problem will intensify in the coming decades as antibiotics are used indiscriminately and become more available globally. Still, this cannot be seen as a proof of macroevolution. Such changes, although profound and sometimes deadly, did not change the tuberculosis bacterium into anything but a more resistant tuberculosis bacterium.

As anyone can clearly see, such observations support only microevolution or minor changes within the originally created kinds. Such is the evidence from the development of antibiotic resistant bacteria for this, too, has failed to provide evidence for macroevolution or the changing of one major kind of organism to another kind. Yet evolution depends on this important, but unseen and unsupported foundational principle. The very foundation of evolution is cracked and the cracks are growing ever wider as more of the details of the profound complexity of all life are discovered. Solid support is lacking after 150 years of searching, by tens of thousands of evolutionists, which explains why thousands of learned scientists and are abandoning evolution in droves. They no longer see it as supported by the scientific evidence. It baffles me how our news media miss or suppress this important trend in science.

Selective Breeding

Charles Darwin spent many pages in The Origin of Species documenting changes in domestic farm animals brought about by centuries of selective breeding. Yet, a sheep remains a sheep and cows remain cows. No new species has been developed. Even the recently cloned animals remain true to their parent species. We understand much more today about genetics than was known in Darwin's day. In fact, it is argued that if the scientific community had known then what we now know about DNA and the ultra conservative way inherited traits are passed from one generation to the next, Darwin's theory of evolution would have been rejected…even laughed at, as becoming the case today.

We now clearly understand that a finite number of genes govern any one trait such as milk or wool production. Once all those genes are selected, additional genetic improvement is not possible. A cow cannot become merely a milk-producing machine or a sheep just a walking wool factory. They must also be able to move about, eat, sleep and reproduce. Here again, Darwin was obviously wrong. He thought genetic changes could continue without limit. This is supported by the then popular argument for the origin of the long necks of giraffes. Darwin and others accepted that by stretching the neck to reach ever higher, this longer neck trait could be passed on to the offspring. Of course, we know today that there is no mechanism by which this could be passed to the offspring. This popular notion was the result of the earlier work of the French biologist Jean-Baptiste Lamarck and was called "soft inheritance" or more commonly "the inheritance of acquired characteristics." Perhaps the strongest proof is seen in Jewish boys. For thousands of years Jewish boys have been circumcised, yet the amount of foreskin of Jewish boys at birth remains undiminished.

Darwin provided many examples he thought clearly demonstrated the inheritance of acquired characteristics. He called this Lamarckian hypothesis Pangenesis and explained it in some detail in the last chapter of his book, Variation in Plants and Animals under Domestication (Darwin, 1868). He defined Pangenesis as a hypothesis based on the principle that somatic cells would throw off microscopic particles called "gemmules" in response to the environment. These gemmules would eventually make their way to the germ cells where they could pass the information about the parent cells to the next generation. Of course, no mechanism for such inheritance has been found and this idea was rejected long ago. The very inheritance Darwin proposed to power evolution prohibits change without limit. Indeed, the opposite trend is now slowly spreading in agriculture with the cloning of livestock there is the total cessation of additional genetic change.

There you have it. That is all the direct evidence there is supporting macroevolution. It is surprisingly weak and inconclusive. Alternative explanations abound. There is evidence supporting microevolution, but for the all-important proof for the origin of life or macroevolution, the evidence is totally missing. It does not exist and never has. The entire superstructure of macroevolution is based on faith and hope and not on any scientific evidence. Since this is the best direct evidence supporting evolution, one can see why doubts are growing today. In a recent poll of qualified scientists with terminal degrees, over fifteen thousand have doubts about evolution…and that number is growing explosively.

As a student, I found it instructive that when biology instructors were asked for proofs that evolution is still occurring, they quickly asserted that the process is much too slow to witness. Yet, when students asked why there is no fossil record of transitional plants or animals changing into something different, we were told that evolution occurs too rapidly to leave fossil evidence. Evolutionists cannot have it both ways. They must openly confess there is very little actual evidence supporting evolution and that it is accepted by faith.

Our students must be told the truth. We should demand nothing less both in our public schools and especially in higher education. Evolution is more faith based than science based and again at the core its acceptance is a moral issue. Yet, it seems the leaders in modern evolution are highly successful in keeping this secret out of textbooks and far from science classrooms. The media seems to be partners in this charade and blatant disregard for truth. Where are those aggressive investigative reporters when you need them? Our culture has been deceived into believing the groundless lie of evolution. Evolution today has truly become a dogma. The reason for the widespread acceptance of this lie is clearly given in scripture. *For this reason God sends them a powerful delusion so that they will believe the lie* (II Th 2:11, NIV). How else can the acceptance of evolution without any hard evidence be explained?

Indirect Evidence for Evolution

The above direct evidences of evolution were not very convincing. Now let us consider the indirect evidences used in most biology textbooks to support evolution. As mentioned earlier, alternate interpretations can be made for these indirect evidences, but they are never mentioned in textbooks or classroom lectures. Instead, they are presented as factual evidences of evolution. Again, we must demand honesty and full disclosure. There is no place for deception in public school and university classes…yet it continues unabated, and is supported by the media. It is time we demanded the truth be taught in our science classrooms. Open discussion on both sides of this important issue is needed both in our public schools and especially in our universities. Only by considering the alternatives can true science make progress. Students and scientists must again have the freedom to follow the evidence no matter where it leads. Such discussion could also sharpen a student's ability in critical thinking.

Evidence From Comparative Anatomy

Let's begin this portion of the discussion with the evidence from comparative anatomy. It is widely used as a "proof "of evolution and is easy to comprehend. It is also simple to see how the same information can also be used just as effectively support the opposing view.

In many university biology programs, comparative anatomy is a difficult, but important and required course for biology majors. It is also the first course where most biology majors begin to accept evolution as established face. Here, students can see the similarity in structure and function of organs and structures of many different animals. Students often dissect several vertebrate animals during the course including a shark, salamander, frog and a fetal pig or cat. Many similarities are obvious and students are taught repeatedly that such similarity of structure is proof that the animals are descended from a common ancestor. Sadly, few students have the knowledge, background or courage to question the evolutionary interpretation touted relentlessly by their professors and textbooks. Most accept the arguments and are won over to the evolution camp for the rest of their lives. Such a view raises lingering doubts for Christian students about what they were taught at home and in Sunday school. Biblical inerrancy is called into question. These doubts linger and grow with each new evolution based course. For non-Christian students, such arguments often erect barriers and block all roads to the Cross and Salvation.

First, we will examine the information presented in lectures and textbooks and see how they can be used to support evolution. Most textbooks use one or more illustrations showing the bones of the human arm, leg of some common mammal such as a dog, horse or cow, the flipper of a whale, and wing of a bird or bat. Indeed, there is a striking similarity in the number and name of the bones and to some degree even in the shape of the bones.

As a side issue, human anatomy was extensively studied and the various organs and structures named long before the internal anatomy of other animals were thoroughly examined. As various animals were dissected and the organs and structures named, it was logical to use the names given to human parts.

201

Therefore, there is a bit of circular reasoning involved when a professor today not only points out the similar structures in various animals, but also uses even the names of the structures as evidence of common ancestry. They seem to forget the anatomical parts were named to show the similarity with humans.

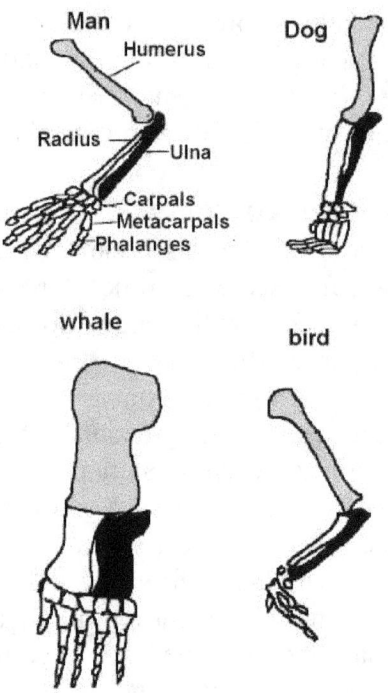

Evidence of Evolution from Comparative Anatomy

As seen in the drawing above, there are similarities in the structure of the forelimb in various vertebrates. Students must learn the name of each of the bones as well as many of structures on each bone. For beginning biology students, this can be a daunting task. As they study skeletal material, they see certain similarities in the way the bones are shaped and especially how they are connected together. Once students grasp the similarity of the bones in the various animals, it is then used as a powerful "proof" of common ancestry…of ameba to human macroevolution. The professor repeatedly uses this anatomical information to illustrate the fish to-amphibian-to-reptile-to-mammal-to-human evolution. They hear the same evolution scheme repeatedly in each biology class, by different professors using similar arguments. It can be overwhelming.

The evidence from comparative anatomy is easy to grasp. For many students, this is the most powerful evidence for evolution they have seen and they understand it. Until the course in comparative anatomy, much of the support for evolution was theoretical. This course makes the evidence for macroevolution more tangible and real. Students hear the same arguments from other professors in other classes and conclude they cannot all be wrong. Indeed, most biology graduates fully accept evolution as the only scientific explanation for the origin of all the plant and animal species and for the origin of man. There is no longer need or room for God as Creator. For many, this is in sharp contrast to what they have been taught at home and at church. It is also very different from the account of Creation in the Bible. Still, as science majors, many began to accept evolution and reject the truth of God's word. I saw this countless times for Christian science majors…and it broke my heart. Few students have the intellectual background, motivation or time to understand that the same structures can also be interpreted in a way that supports biblical Creation, as we shall see below.

The argument from comparative anatomy is only indirect evidence and certainly not an actual proof that evolution occurred. Yet, it is dogmatically stated by science professors as proof positive to inexperienced and often naïve students. It can just as easily be seen to support an all-wise Creator. Perhaps here pastors and Sunday school teachers should be better prepared. The following illustrates how easy it is to view the same evidence from comparative anatomy as supporting Creation instead of evolution.

Certainly, comparing the anatomy of various vertebrates can be seen as evidence of common ancestry. The evidence can, however, be seen as supporting the creation worldview with the same certainty. Animals contain a finite number of bones and other organs. To provide movement and mobility, some of these bones are used in appendages in the arms and legs. It is difficult to imagine how each structure could be totally different and unique. God, in His infinite wisdom, used similar structures in a wide variety of animals to provide movement. It is just as logical to see the similarity of structures in comparative animal anatomy as evidence of a common designer as it is to see evidence of a common ancestor. Again, the forelimb of whales, cattle and even birds are all used for locomotion and can be seen as evidence of a common Creator.

A simply analogy will illustrate this important point. Assume a famous architect designs a series of different buildings. One is a home; another is a convenience store, a bank and a tall skyscraper. Each has different functions but certain constraints exist regarding the availability of building material and common conditions such as wind, rain and gravity. No doubt, the mark of the architect would be clearly seen in each building, even though they were designed for different functions. It is the same with the anatomy of the various animals. The mark of the Creator can be clearly seen in each of them. What some see as evidence of evolution and common ancestry; others see as evidence of a common Creator. Both views fit the evidence and are equally plausible. Neither is a proof positive and both are dependent on one's worldview. Our students must be taught to see both viewpoints regarding the evidence from comparative anatomy. Let's examine another indirect evidence of evolution. This one is also fun because like with the infamous peppered moth, the comic duo of fraud and deception make another visit.

Evidence From Comparative Embryology

Embryology is the portion of biology that deals with the embryo and its development from fertilization to birth or hatching. Eventually, every parent hears the dreaded question from their child, "Where do babies come from?" Embryology attempts to answer that ubiquitous question in depth. This is perhaps the oldest and best known argument used as evidence to support evolution. It is also the weakest. The concept is introduced with a series of drawings presented as the embryonic stages of various animals including man. Most biology textbooks contain some form of the drawings shown below. Notice how the early embryos look alike, but become quite different as they grow and mature.

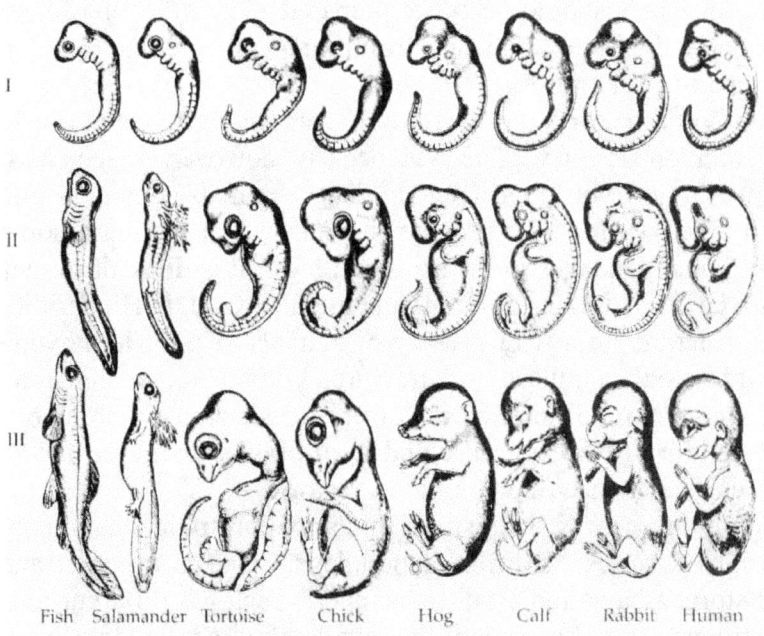

Fraudulent Embryo Drawings Credited to Ernst Haeckel.

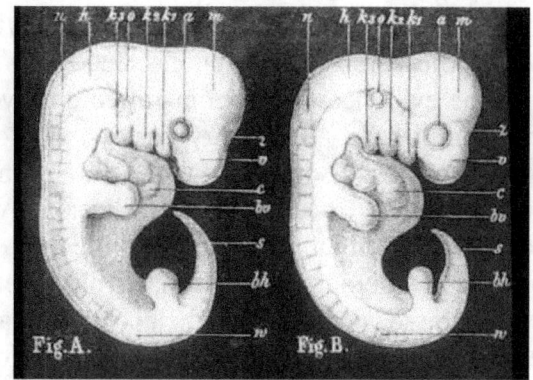

Early Human and Dog Embryo from Haeckel.

I must confess, the first time I saw the drawings of various early embryos I felt intimidated and frightened. I assumed the drawings were factual; after all, they were in my biology textbook. I trusted my professors would not use a textbook with fraudulent information. I was deceived, as were my professors. Below are the actual embryos Haeckel claimed to have used for the above drawing.

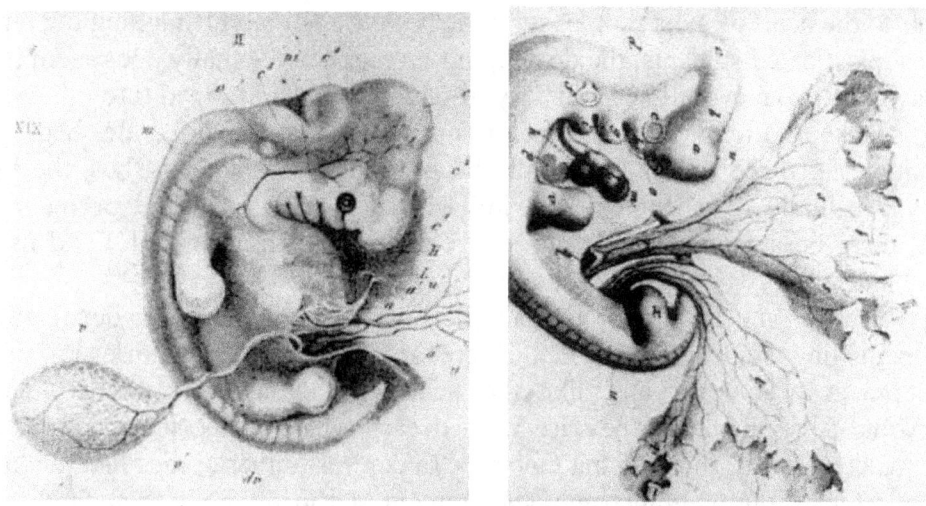

Actual Human and Dog Embryos Haeckel Said He Copied.

Notice how very different the two embryos actually look. He obviously cheated and made them look more alike. He was caught cheating, tried in a court of law, found guilty and finally confessed to the fraudulent drawings. However, for decades, these known fake photos have been used to brainwash our students. This is simply not right and must be stopped. Where the evidence is weak it must be admitted, but there is absolutely no excuse for using known false evidence to support evolution dogma. Again, what happened to objectivity in science? Do scientists and university professors not have a conscience?

There is more. The evidence from comparative embryology is steeped in those beautiful words familiar to all biology majors, "ontogeny recapitulates phylogeny." It has such a beautiful sound and rolls off the tongue so effortlessly. Many biology majors find a way to drop it into dinner conversation with their parents. A typical conversation might go something like this. "Mom, please pass the potatoes and did you know ontogeny recapitulates phylogeny?" Unsuspecting parents are often so impressed with what their children are learning at the university that more spending money or a new car may be forthcoming. Those useful words simply mean that during embryologic development (ontogeny) an embryo replays (or recapitulates) its own evolutionary development (phylogeny). This concept is also referred to as the Biogenic Law adding even more weight to the argument.

It is true the human heart starts beating when the baby is only two and a half weeks old. It is also true that the human heart at this stage consists of little more than a contracting tube, not unlike the heart of an adult earthworm. The human heart then progresses through stages in which it resembles the hearts of fish and amphibians. Certainly, prior to birth, the human heart is remarkably similar to the heart of most reptiles with the bulk of the blood bypassing the lungs. To unsophisticated students, the developing human heart actually does seem to be portraying its own evolutionary history in which it has developed through worm-like, fish-like and reptile-like stages. It is also true that all vertebrates begin life as a single fertilized egg and gradually become miniature adults. Thus, the early embryos actually do show certain similarities be they fish, bird or mammal. Could there be a more obvious proof of evolution? Ah, a beautiful and useful phrase indeed.

There is much more to this simple portrayal of embryonic development than beginning students are told. There are two broad categories of organ development. Some organs, like the heart and kidneys are needed during embryonic development. They are formed early and become more complex to meet the demands of the growing embryo. In contrast, other organs like the lungs are not required during embryonic development and form late in their final form. According to the Biogenic law, this would imply we evolved from animals that had no lungs. The sex organs are not needed until maturity and are not fully developed at birth. Again, the logical conclusion would be that we evolved recently from animals that lacked sex organs. Obviously, the whole argument from comparative embryology falls apart. God designed animals to develop in a logical and functional way. Organs not needed until birth or after develop late and in final form. This simple explanation can account for the similarities of certain structures during embryonic development, yet can also account for the differences.

Evidence from Vestigial Organs

This has long been another of my favorite the evidences for evolution, for it is easily dismissed and once again illustrates the deception common in textbooks and biology classes. This argument is also easy to comprehend. If animals actually evolved over time from one major kind of animal to another, we would expect to see some organs no longer needed as well as nascent organs or organs on the verge of becoming useful. No nascent organ has ever been described, but the argument of vestigial organs is still widely used to prop up evolution.

Robert Wiedersheim (1848-1923) was not a good student, barely passing his final exam and his academic advancement was slow. In 1876, he became an anatomist at the University of Freiburg and soon became a comparative anatomy expert publishing several textbooks. He became widely known and respected in 1893 for publishing a list of 86 useless or vestigial organs (Wiedersheim, 1893). In his own words, he said each of those human organs had, "lost their original physiological significance." He theorized they were vestiges of past human evolution and called them "vestigial." Over the next few decades, he and others added additional organs and the list grew to 100 and then to 180 useless organs by the time of the infamous Scopes Monkey Trial. I searched diligently for the actual list of 180 organs mentioned in many textbooks, but was unsuccessful in finding such a list. This seems to be yet another myth started and widely disseminated by evolutionists.

Wiedersheim claimed the human body was a veritable walking museum of evolutionary history. He picked up and expanded on Darwin's concept of rudimentary organs as mentioned in *The Descent of Man*. Original organs included on Wiedersheim's list as useless included the human appendix, adenoids, tonsils, parathyroid gland, pineal gland, pituitary gland, thymus, valves in veins and many other important organs. As knowledge replaced ignorance, uses were found for these organs. Today, no one would claim any useless organs exist in the human body.

Perhaps best known of these is the appendix. For decades, medical doctors actually accepted the evolutionary myth that the appendix was useless and during other surgical procedures, even healthy appendixes were removed, increasing surgical complications and prolonging recovery. Doctors also profited from the needless surgery.

The human appendix is a finger sized hollow tube near the end of the small intestine and entrance of the colon, and has long been taught to be a relic of our vegetarian ancestors and no longer serves a function today. Today, we know it serves many functions. For example, over fifty years ago we find these words from the prestigious *Quarterly Review of Biology*, "There is no longer any justification for regarding the vermiform appendix as a vestigial structure" (Straus, 1947). As a blind reservoir, it continually repopulates the colon with important bacteria lost in excrement. The walls of the appendix contain lymph tissue, known to be important in the immune response to disease. Recent evidence indicates it also lubricates oil needed for lubrication of the large intestine. Yes, I can live without my appendix or without my right arm, but this does not mean God did not give them to me for a purpose.

So it is with each of the so-called vestigial organs on the original list published by Wiedersheim. Yet, this argument is often still heard in classrooms and is taught in the older textbooks. Again, evolutionists must cling to yet another myth because the actual evidence supporting evolution is non-existent. Let us look briefly at some of those other organs originally listed as useless.

We now know the adenoids and tonsils contain lymph tissue and help in our immune response to disease. The pituitary gland, once classified as useless, is considered the "master gland of the body" for it influences virtually every biological pathway throughout the human body. By way of the important hypothalamic-hypophysial portal system connecting our brain and body, emotions can have a profound influence on our bodily functions.

The parathyroid is vital in the regulation of calcium levels and maintaining healthy bones. The thymus plays a key role in the immune response and is vital. Even the wisdom teeth have been shown necessary for proper development of the jawbone. It has long been known valves in the great veins of the body are important in returning blood to the heart. Another one that often appears on vestigial organ lists is the so-called third eyelid or nictitating membrane we all have. I don't know about yours, but mine is functional. Often, when I awake there is some sticky gunk attached to mine. My nictitating membrane functions as a sticky gunk collector and is therefore not vestigial. It should be obvious to all there are no useless organs…only human ignorance. Our Creator knew precisely what He was doing when He made our marvelous bodies. Only by willful ignorance can people deny the obvious truth of the Psalmist, *For you created my inmost being; you knit me together in my mother's womb. I praise you because I am fearfully and wonderfully made; your works are wonderful, I know that full well.* (Ps 139:13-14, NIV) Indeed, *The fool says in his heart, "There is no God." They are corrupt, their deeds are vile; there is no one who does good.* (Ps 14:1, NIV)

Evidence from Taxonomy

This is perhaps the weakest of the evidences used to prop up evolution because it contains a huge element of circular reasoning. Taxonomy or systematics is the orderly scientific classification of plants and animals. Each plant or animal has a binomial name consisting of the genus and species. Genera are grouped into families, families into orders, orders into classes, classes into phyla and phyla into kingdoms. Perhaps an example will make it easier to comprehend. My favorite animal is the American alligator known scientifically as *Alligator mississippiensis*. Because it is Latin, it is written in italics and consists of the genus, *Alligator* and the species name *mississippiensis*. Together they make up the scientific name of the alligator common in the southeastern United States and the object of my study for decades. The only other living member of the alligator genus is the Chinese alligator, *Alligator sinensis.*

Alligators are classified in the family Alligatoridae, again with only two living alligator species. Alligators are classified in the order Crocodilia including alligators, crocodiles, caiman and a few other less known members. They belong to the class Reptilia sometimes called Sauropsida including all reptiles. This class includes not only crocodilians, but also turtles, snakes and lizards. Reptiles belong to the phylum Chordata including all animals with backbones. This also includes fish, birds and mammals. The chordates belong to the animal kingdom or Animalia. The number of kingdoms seems to be in dispute and includes the animal kingdom, plant kingdom and three other lesser known kingdoms to account for bacteria, fungi and some other organisms. The details of taxonomy are beyond the scope of this discussion and readers interested in knowing more can readily find more information in any library and many introductory biology textbooks.

Modern taxonomy is an attempt to arrange living things in a way that shows their alleged evolutionary relationships. To try to show the taxonomic relation of various animals and plants as a proof of evolution is circular reasoning because they were purposely arranged to show such a relationship. Even as a student, I saw this evidence as weak and still think it is laughable for textbooks and professors to teach it as evidence supporting evolution.

It has been argued that arranging things by similarity is a very human trait. Here is an often used example. Assume you are shipwrecked on a tropical island with ample food and water. The ship was filled with miscellaneous hardware...bolts, nuts washers and such. These items were somehow beached at the time of the shipwreck, but randomly mixed Out of sheer boredom, many of would find ourselves going through the hardware and sorting it by size and function. Such sorting and arrangement seems very human. It is a way to find order in chaos.

So it is with the multitude of living organisms. Taxonomy provides such order, but is in no way a proof of common ancestry any more than would be a completely organized grouping of hardware implies each kind arose from a simpler bolt, nut or washer. The argument from taxonomy has absolutely no merit as an evidence of evolution.

There are a few other lesser arguments often used to support evolution. Again, there are problems with each of these. Let's end this discussion of the evidences used to support evolution with the single most important evidence, that of the fossil record. Some professors and many textbooks admit much of the preceding evidence is circumstantial at best and other interpretations are possible. Many say for the actual proof evolution, one must turn to the fossil record, the actual history of life on planet earth. Does the fossil record support evolution?

Evolution and the Fossil Record

This is a very important topic for evolutionists. Many biology textbooks adamantly teach the fossil record proves that evolution has occurred. If there is indeed fossil evidence supporting evolution, then it should be taught in high school and university biology classrooms. If instead, there is little actual support for macroevolution, this must be clearly stated. The actual fossil record is often ambiguous, supporting alternate views, yet, even the discussion of those problems and contradictions are banned from university classrooms. Let's consider the fossil evidence in some detail.

Let me begin with a true story that will set the tone for this discussion. For decades, Duaine Gish, Henry Morris, Sr. and other Creationists openly debated evolutionists about the evidence supporting evolution. Many years ago, Dr. Gish debated a leading evolutionist in England. In preparation for the debate, the evolutionist he was debating went to the British Museum of Natural History for the latest evidence. This is a world class museum with one of the world's best collections of fossils. The evolutionist met with the fossil curator and asked for the best fossil evidence he could use to blow Dr. Gish out of the water. The curator of fossils at the British Museum of Natural History sternly warned the evolutionist to stay far away from fossils for in truth they fail miserable to support evolution. I have long found this advice from someone who knows, most revealing, for it flies in the face of what our textbooks and professors teach students. It remains as true today as it was then…the fossil record does not support evolution.

The modern evolutionary synthesis was put together in the 1930s and 1940s by Theodosius Dobzhansky, Ernst Mayr, J.B.S. Haldane, Sewall Wright, George Gaylord Simpson, G. Ledyard Stebbins and others. But as this philosophy was fine-tuned and disseminated to the masses, it was never actually observed from the fossil evidence. I was shocked and angered when I discovered this important fact. Generations of students have been deceived by the education system. ***Phyletic gradualism [gradual evolution] was an a priori assertion from*** *the start – it was never "seen" in the rocks* (Gould & Eldredge, 1977) yet is presented as absolute fact and proof of macroevolution in countless high school and university biology textbooks and by the teachers in biology and other classes. When informed students challenge the aforementioned weak evidences supporting evolution, the professors reflexedly respond that for the real proof of evolution, one must only look at the fossil record. The same argument is often used in textbooks. This is central to the acceptance of evolution. Does the fossil record clearly support evolution as it is adamantly reported to do? As we shall clearly see, the fossil record also fails to support macroevolution. Instead, the rich fossil record is proof positive of a global flood and rapid burial. Let's consider the fossil evidence in some detail.

The history of most fossil species includes two features particularly inconsistent with gradualism (neo-Darwinism) and their sudden appearance: stasis and sudden appearance. *Most species exhibit no directional change during their tenure on earth. They appear in the fossil record looking pretty much the same as when they disappear; morphological change is usually limited and directionless. Sudden appearance. In any local area, a species does not arise gradually by the steady transformation of its ancestors; it appears all at once and "fully formed."* (Gould, 1980)

Gould was correct. Lack of fossil evidence supporting Darwin's idea is repeatedly admitted in secular literature. Consider the following: *Both the origin of life and the origin of the major groups of animals remain unknown* (Fisher, 2003). *Almost all of our information about macroevolution has come from the fossil record and, as we have seen, there are doubts as to how reliable this is* (Palmer, 1999). *Perhaps no aspect of evolution has received such intense study as human evolution, yet this is a subject concerning which there is much debate, and about which there is still much to be learned* (Colbert, 2001). Such statements by learned secular scientists has helped the cause of ID/creation science, for non-evolutionists have been saying this since Darwin.

Skeptics of Darwin's theory have used a truly remarkable book by evolutionist Barbara J. Stahl of Saint Anselm College in New Hampshire, revealingly titled, *Vertebrate History: Problems in Evolution.* Sadly, this important work is out of print. Dr. Stahl, anatomy professor and paleoichthyologist, is clearly no friend of the creationist. She was, however, intellectually honest enough to write this 604-page book documenting the many problems associated with alleged evolution of the vertebrates. Darwinists were understandably quick to downplay Dr. Stahl's research. In recent years, their only "valid" criticism is that the book is dated and anything found in its pages are now passé. I strongly disagree. In 2001, Edwin H. Colbert and his coauthors published their fifth edition of *Colbert's Evolution of the Vertebrates*. Dr. Stahl's detailed research has held up all these years when compared with Colbert's more recent text. Consider carefully the following examples from her book.

Origin of Fish: "The higher fishes, when they appear in the Devonian period, have already acquired the characteristics that identify them as belonging to one or another of the major assemblages of bony or cartilaginous forms" (Stahl, 126). "Both these groups appeared in the late Silurian period, and it is possible that they may have originated at some earlier time, although there is no fossil evidence to prove this" (Colbert, 53). Contrast this lack of fossil evidence for evolution with the clear evidence for creation: the sudden appearance of fully formed vertebrates (and invertebrates!) in the fossil record. Is it any wonder that another honest evolutionist stated, *The origin of animals is almost as much a mystery as the origin of life itself* (Donoghue, 2007).

Origin of Amphibians: Since the fossil material provides no evidence of other aspects of the transformation from fish to tetrapod, paleontologists have had to speculate how legs and aerial breathing evolved (Stahl, 195). This is certainly a logical explanation of the first stages in the change from an aquatic to a terrestrial mode of life. We can only speculate about this. (Colbert, 84-85).

Origin of Snakes: The origin of the snakes is still an unsolved problem" (Stahl, 318). Unfortunately, the fossil history of the snakes is very fragmentary, so that it is necessary to infer much of their evolution. (Colbert, 154).

Origin of Birds: In the absence of fossil evidence, paleontologists can say little about the date at which these sixty-nine living families of Passeriformes appeared. (Stahl, 386) Of all the classes of vertebrates, the birds are least known from their fossil record. (Colbert, 236).

Origin of Whales: As with most tetrapods secondarily modified for aquatic living, ascertaining the terrestrial stock from which the whales came is exceedingly difficult. (Stahl, 486). Like the bats, the whales (using this term in a general and inclusive sense) appear suddenly in early Tertiary times, fully adapted by profound modifications. (Colbert, 392). Many years ago, I had the following somewhat whimsical drawings made to help emphasize this point in my Creation lectures.

A Whale of a Tale
        There are no fossils linking whales to any other group of vertebrates. Undaunted evolutionists are confident whales evolved from some unknown

212

ungulate. They would have us believe some cow-like ungulate ancestor of the whale made its way to the ocean for a refreshing swim.

She seemed to enjoy the water and gradually ventured farther out to sea. Her legs were mysteriously becoming more useful as fins gliding her gleefully along.

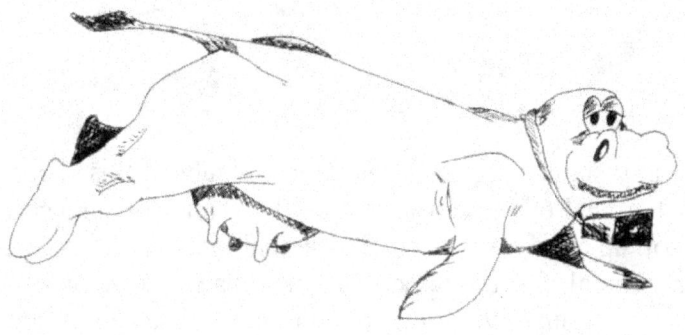

As if by magic, her legs were being modified into well designed flippers and she learned how to hold her breath longer and longer. All the time she was very careful not to leave any fossil record of her slow transformation.

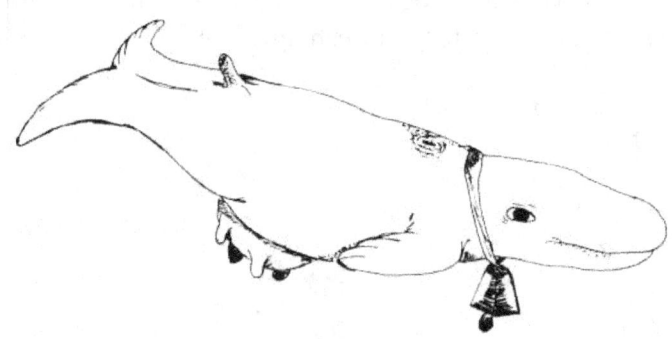

Slowly, ever so slowly, over millions of years as if by some intelligent design, she was magically transformed into a modern whale. Again, making certain no fossil evidence was left in the ubiquitous sedimentary deposits.

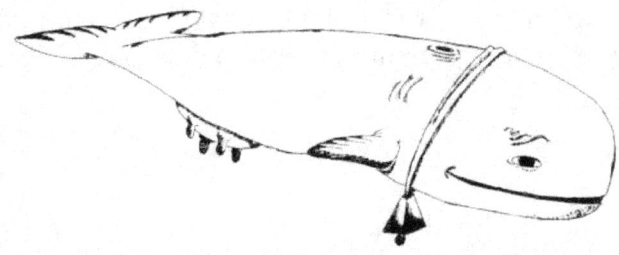

Lacking any supporting fossil evidence, I submit this theory is an udder failure, yet this fictional tale is repeated countless times in science classrooms in an attempt to prop up a failing theory. The actual facts from the fossil record and from science in general, fail to support gradual change over time, yet that is the very essence of evolution. Evolution is becoming increasingly bankrupt for lack of any actual supporting scientific evidence. Once again, I say vehemently it is the evolutionist and not the Creationist that has the greater portion of blind faith for I *know* in whom I believe.

Evolution Acceptance

Darwinism soon arrived on the shores of the United States, causing a political upheaval that climaxed in the infamous Scopes Trial in Dayton, Tennessee in 1925. It is important for Christians and others to understand one of several pivotal points about this historical event in American history. The newly-formed ACLU sought a teacher who would test the Tennessee Butler Act that forbade evolutionary instruction. Soon, a young substitute teacher (who was not even trained as a biology teacher) named John Scopes was accused of teaching Darwin's theory in the Dayton school. This delighted the secular community, especially the local businessmen who saw this as an opportunity to put Dayton on the map. Evolutionists were waiting for such an event to happen so the law could be tested (the Tennessee Butler Act was eventually repealed in 1967). This particular origins foray pitted the Biblical account of creation with the secular, no creatorless, molecules-to-man philosophy. Although the creationists technically won the case, it was a hollow victory.

The reason Darwin is so revered by the secular world today is he is thought by many to be the first to provide a plausible explanation for the origin of life, diversity of living organisms including origin of man without the need for a Creator-God. As is shown above, this premise is clearly false for there is overwhelming evidence of Creation. Indeed, as King David of old said: *The fool says in his heart, "There is no God"* (Ps 14:1a, NIV).

Reasons For The Acceptance Of Evolution

Contrary to what most people think, evolution is not accepted because of overwhelming supporting scientific evidence. As previously stated, the acceptance of evolution was, and is, at the core a moral issue. This is an important point, yet is not widely known. Let us examine this important premise in some detail.

Science has given our modern world many things and scientists are respected as few others. It seems people in our society today accept whatever scientists say even in matters outside their areas of expertise. Many feel they must accept evolution because they have been told repeatedly this is what all scientists accept as true. It is always taught as absolute established fact. Many Christians then attempt to see evolution as the method by which God created living things. This view is even taught in many Christian schools and universities. Could this be true? Did God use evolution to create the living world? Scripture is very clear. Creation was a miracle. It was not a process. All kinds of living things were created during the six literal day week of Creation. The days of Creation in Genesis are clearly defined as literal twenty-four hour days with the repeated phrase, "evening and morning." There is more.

Leading evolutionists always define evolution as a fully natural process. In other words, if God or any unidentified intelligent designer played a role in the process, it is not evolution. Many Christians think that evolution has only modified the scientific view of the supernatural Creation. Nothing could be farther from the truth! Evolution totally excludes any form of supernatural involvement in

Creation. Let us next consider what some of the leading scientists believe about God and evolution.

From a college Botany textbook: In the years following Darwin's Origin of the species (1859), the theory of evolution gradually REPLACED the concept of special Creation. There is no room for God or creation in modern science. Isaac Asimov was one of the world's most influential scientists. He and Carl Sagen strongly influenced people's view of science. Dr. Asimov said regarding origins, In the beginning how did life begin? It seems quite certain that life developed. NOT AS A MIRACLE, but merely because molecules combined with each other along the line of least resistance. Life could not help forming under the conditions of the primitive earth any more than iron can help rusting in moist air. Notice once again, how God is totally excluded from Creation. God must not even be seen even as a First Cause. Again, I cannot over-emphasis that evolution is always defined as fully natural or atheistic process. If God played any role in it, then it is simply not evolution. Those seeing God as directing evolution have neither the support of science nor the Bible. They stand alone.

The next quotation from the late British Biologist Sir Julian Huxley is more to the point: In the evolutionary pattern o thought there is no longer need or room for the supernatural. The earth was NOT created: it EVOLVED. So did all the animals and plants that inhabit it; including our human selves, mind and soul, as well as brain and body. So did religion. There is no doubt, where Professor Huxley stands in his relation to God. We should thin twice before we let his view of origins influence our interpretation of God's Word.

Next, we hear from the world-renowned American evolutionist of Harvard University, George G. Simpson. He was president of one of the world's foremost scientific organizations. I was actually at the international meeting when he said the following: *Evolution is a fully natural process. Inherent in the physical properties of the universe by which life arose in the first place and by which all living things, past or present, have since developed. Organisms diversify into literally millions of species then the vast majority of those species perish and other millions take their places for eon until they too are replaced. If that is a foreordained plan, it is an oddly ineffective one...* What Professor Simpson is saying is that if this is the way God created living things, then God was stupid. He goes on to say, *A world in which man must rely on himself... is by no means congenial to the immature or the wishful thinkers...Life may conceivably be happier for some people in the other worlds of superstition. It is possible that some children are made happy by a belief in Santa Claus. But adults should prefer to live in a world of reality and reason.* The pattern could not be clearer. The leaders in modern evolution always exclude God. There is neither room nor need for the supernatural. To them life, including man is nothing more than matter in motion. Any need for God has been replaced by their view of science.

216

Why evolution was so quickly and widely accepted is an important question. Many educated people would say the evidence amassed by Darwin was overwhelming. Not so. Morality was, and is, at the core of the creation/evolution debate. Sir Julian Huxley was, until his death, the world leader in modern evolution. When asked why the scientific community so quickly embraced evolution, he did not mention the evidence. Instead, he said, *The reason we leapt at evolution was the idea of God interfered with our sexual mores.* So it is today, acceptance of evolution removes God from creation and from conscience. Again, we see the acceptance of evolution is a moral issue, not science.

Another famous Huxley from Great Britain, Thomas Huxley, Grandson of Aldous Huxley, is one of Darwin's strongest supporters. He has been one of the most influential evolutionists, perhaps in some ways even more so than Darwin. His publications are often cited in university classrooms today. According to Huxley, *I had motives for not wanting the world to have meaning. The philosopher who finds no meaning in the world is not concerned exclusively with a problem in pure metaphysics: he is also concerned to prove that there is no valid reason why he personally should not do as he wants to do--The philosophy of meaningless was essentially an instrument of liberation. The liberation we desired was liberation from a certain system of morality. We objected to the morality because it interfered with our sexual freedom.*

Let's conclude this section with what the Bible teaches. God's Word says it best. ***There is a way that seems right to a man, but its end is the way of death.*** (Prov.14:12) And, ***The fool has said in his heart. "There is no God."*** (Psalms 14:1) Many leading scientists today have indeed become foolish in their unrelenting denial of God as Creator in spite of overwhelming evidence to the contrary.

Unlike the author, not all are enchanted by alligators.

Chapter 8
Evolution in the Twenty-First Century

Although they claimed to be wise, they became fools. (Rom 1:22)

The United States was founded in large part by Christians wanting to establish a country that would have religious freedom for themselves and assure it for future generations. Religious intolerance caused the Pilgrims to leave England, first migrating to the Netherlands and finally to America founding Plymouth Colony in 1620. Their plight had a profound influence on future American law regarding religious freedom. This is evidenced by the establishment clause in the Bill of Rights that reads: *Congress shall make no law respecting an establishment of religion, or prohibiting the free exercise thereof; or abridging the freedom of speech, or of the press; or the right of the people peaceably to assemble, and to petition the Government for a redress of grievances.* What was intended by these carefully chosen words should be obvious to all. The founding fathers wanted to assure freedom from any form of government dictated religion. They wanted citizens of the newly formed United States to be free to practice their faith in any way they choose.

Changing times

How times have changed with the passing of 390 years. Unfortunately, those powerful words are blatantly misinterpreted today as meaning freedom from religion, especially the Christian faith. Because of this travesty, the Bible and Ten Commandments have been removed from public schools and many public places. Prayer, the mention of Jesus or the Christian God is strictly forbidden in our public schools and public events. A growing movement is attempting to ban the Bible from society. A military chaplain was court marshaled for praying in the name of Jesus. Public school teachers can no longer teach ethics or morality to our children. Yet, unbelievably these same students are now forced to recite Muslim prayers and learn the five pillars of Islam.

What has happened to the separation of religion in our government schools? It seems the only religion deemed illegal by the courts and liberal media today is Christianity. With the ever-popular Harry Potter books, witchcraft is considered an acceptable religion suitable for class discussion in grade schools. Internet assignments are given that include dangerous witchcraft and occultist websites. Christianity is no longer the predominant religion of the United States. It has been replaced by secular humanism as well as a variety of other religious practices including witchcraft, Buddhism and the Muslim faith. Christianity is increasingly slandered or mentioned only in a disparaging tone. Just a century ago, pastors were revered in the community. Today, in popular television programs, pastors are often shallow buffoons or twisted villains. Christians are considered ignorant superstitious fools. A social science professor traveling from a distant planet and studying the present culture of the United States today could only conclude secular humanism is the state sponsored religion and Christianity a forbidden cult.

Most scholars agree naturalistic evolution is the foundation on which secular humanism rests. It is therefore important to examine the history and supporting evidence for such an important idea now considered to be at the core of both science and religion. As we have seen, the actual scientific evidence supporting naturalistic macroevolution is virtually non-existent. Instead, evolution is accepted by faith and has become the dominant religion among the self-proclaimed intelligentsia. Charles Darwin is revered today as few others. To many, he has achieved an unholy form of sainthood. During the non-stop celebration of his two-hundredth birthday, there was even talk of making his birthday a public holiday. Let us look closely at this Saint Darwin and the theory attributed to him. Is such extreme adoration justified?

Misconceptions Abound

In spite of the deference afforded Charles Darwin, many misconceptions surround the man and his writings. Perhaps the most pervasive misnomer today is that evolution is presented in biology classes as a theory. It is not. Nothing could be farther from the truth, for it is always presented dogmatically as absolute fact both in textbooks and especially in lectures by learned professors. Many liken evolution to the law of gravity or with the certainty of the earth traveling around the sun. Unlike most areas of life sciences, evolution is treated as a "hard" science such as physics, chemistry or even mathematics. What is striking is, unlike the laws of physics and chemistry, is that there's virtually no solid experiemental support for evolution. In fact, many have admitted it is impossible to design an experiment or series of experiments that could unequivocally falsify evolution. Again, this is in striking contrast to the laws of chemistry and physics for all of them are firmly rooted in experimental science. At any time, a single experiment could invalidate their acceptance. In chemistry and physics, when a theory is tested and found wanting, it is either drastically modified or abandoned completely. Not so with evolution. It stands alone in science as being unproven…even untestable. Something is indeed amiss. Let's look at some of the tainted history leading to the current widespread and blind acceptance of evolution dogma.

Many today believe and teach that Darwin was the first to espouse the concept of evolution or the modification of living organisms through time with the publication of *The Origin of Species* in 1859. This is simply not the case. The concept of evolution had been widespread since the middle of the 1700's. Evolutionary interpretations were increasingly advanced in the second half of the eighteenth and first half of the nineteenth centuries. (Mayr, 1972). Darwin was simply another follower in this vast field, not the first and not certainly not the leader he is touted to be today.

After reading Lyell's *Principle of Geology* and Malthus's *Principles of Population*, it is taught that Darwin took the theory of evolution that already existed and added 'his' idea of natural selection as the mechanism that powered evolution. Indeed, the term is synonymous with Darwinism and became part of the full title of his book, *On the Origin of Species by Means of Natural Selection, or the Preservation of Favoured Races in the Struggle for Life.* One should keep in mind, however, that Scottish botanist and atheist Patrick Matthew (1790-1874), formulated much of what was later credited to Charles Darwin. Matthew described his ideas in the seldom cited *On naval timber and arboriculture, with critical notes on authors who have recently treated the subject of planting,* in 1831 (see also Harris, 1981). Alfred Wallace, also read Malthus, formulated the concept of natural selection at the same time as Darwin. *Wallace independently achieved and set forth the same ideas as Darwin. He was an independent discoverer of natural selection* (Eiseley, 1959). Even though this is known in science circles, Darwin is often given full credit for first describing the importance of natural selection in the origin of new species. Although Darwin summarized his view of evolution in an essay in 1844, it was none other than Darwin's own grandfather, Erasmus Darwin, who formulated much of the foundation of evolution. This seems to be missed today by leading evolutionists and certainly by the popular media including science documentaries seen on "educational" television programs and more pervasively in university classroom discussions.

Erasmus Darwin favored the evolution of animals from *one living filament* and assembled evidence from embryology, comparative anatomy, systematics, geographical distribution and pertaining to man, many of the facts of history and medicine available at the time. These arguments about the transformation of living things were familiar to Charles Darwin. In fact, Erasmus Darwin originated almost every important idea regarding the mechanism of animal transformation that has ever appeared in evolutionary theory. (Darlington, 1959) Darwin is given credit for the idea of evolution, not because it was original with him, but because of the volumous amount of evidence he amassed.

There was another and perhaps more important reason for the rapid and nearly universal acceptance of Darwin's work. As is well known, the first edition of Darwin's book sold out the same day it was published. It did so, not because of the scientific nature of the book (he never even discussed the origin of species!), rather because the spiritual climate of the day was seeking an explanation to get around the Genesis account of creation. Darwin's theory provided an ideal vehicle for removing the Creator from creation and from society. This need to remove God from society is even stronger today. Such has been the desire of natural man all the way back to the Garden of Eden. Man wanted nature without the Creator. Man craved freedom for his own sinful rebellious behavior without consequences. Many scientists in Darwin's day wanted all of nature to be fully natural and devoid of any supernatural being as do human secularists today. Darwin's tome satisfied this philosophical desire (see Desmond & Moore, 1991). Not surprisingly, there are many today who give adulation and praise to this naturalist, calling him the "Newton of biology." Others call Darwin the greatest scientist who ever lived, a title he and others in his day would have found amusing.

In truth, Darwin was not a thinker and he did not originate the ideas that he used. He vacillated, added, retracted, and confused his own traces. As soon as he crossed the dividing line between the realm of events and the realm of theory he became "metaphysical" in the bad sense. His power of drawing out the implications of his theories was at no time very remarkable, but when it came to moral order it disappeared altogether, as that penetrating evolutionist, Nietzsche observed with some disdain (Barzun, 1959).

Perhaps Soren Lovtrup said it best when summarizing Darwin's ideas and efforts. I suppose that nobody will deny that it is a great misfortune if an entire branch of science becomes addicted to a false theory. But this is what has happened in biology: for a long time now people discuss evolutionary problems in a peculiar 'Darwinian' vocabulary – 'adaptation,' 'selection pressure,' 'natural selection,' etc. – thereby believing that they contribute to the explanation of natural events. They do not, and the sooner this is discovered, the sooner we shall be able to make real progress in our understanding of evolution (Lovtrup, 1987).
Clearly, Darwin's ideas were in conflict with objective scientific inquiry and empirical observations then as they are today. He admitted as much in chapter six of his book (1859) entitled 'Difficulties on Theory' and again in chapter nine (first edition) 'On the Imperfections of the Geological Record.' Furthermore, empirical science (e.g. laboratory research) is not the same as historical or origin science. Experimental science can be repeated and theories discarded if found to be unreproducable or in error. Historical science is much more difficult to disprove and often prone to alternate interpretations.

In sharp contrast, Christians have the written record of someone who was there in the beginning, versus the secular scientist who can only look to chance, time and random natural processes. Orthodox Jews have the Old Testament and

Muslims their *Qur'an,* both of which also hold to supernatural origins in opposition to secular science naturalism and Darwinism.

Eclipse of Darwinism

The year 1882 is known as the Eclipse of Darwinism. At that time, the Victorian culture began to undergo a significant transformation and in the scientific field skeptics did not consider Darwin's natural selection to be an adequate explanation of origins. Indeed, Darwin himself understood that selection alone was unable to explain the origin of life itself or of any new species. The actual mechanism of Darwinian change ("descent with modification") was as much in the dark then as it is today. Then why, one may ask, was Darwin popular? The answer is simple: he explained creation without a Creator. The late S.J. Gould said that Darwin's idea of gradual evolution expressed the cultural and political biases of nineteenth century liberalism (Gould & Eldredge, 1977). Evolution was and remains accepted in large part because it removes God from culture and from conscience.

Darwin stated that new traits in an organism came about "from use and disuse, from the direct and indirect actions of the environment." This flawed idea is called pangenesis and became popular four decades before Darwin by a famous Frenchman Jean-Baptiste de Lamarck (1744-1829). But Darwin also believed this inheritance of traits acquired by use and disuse and said so, using giraffes on the ancient African prairies as an example. As generations of short-necked giraffes stretched their necks to get leaves in higher trees, they produced and passed on to the next generation "stretched neck pangenes." This idea has long been shown to be unscientifically sound, although it did make an impact in social Darwinism with the theme of "progress through effort." Actually, there has been very little scientific interest in showing that a giraffe's longer neck is an example of 'evolution by natural selection,' as revealed by two evolutionists from the University of Pretoria (Cameron & du Toit, 2007). Later, some of Darwin's ideas of evolution were combined with Lamarck's theory that resulted in neo-Lamarckism. Still, there is no known mechanism whereby a parent can pass on acquired traits, such as a longer neck achieved by continual stretching to reach tree leaves, to its progeny. A world class weight lifter will not produce genetically stronger children simply because he exercised his muscles and became strong.

In summary, many misconceptions about Darwin and his writings abound. Neither the concept of evolution nor its alleged mechanism were original with him. Many of the ideas he discussed have been proven wrong and others lack any supporting evidence. Still, he has indeed had a major and seemingly lasting influence on biology because people desperately wanted a scientifically defensible way to explain the origin of life and diversity of species apart from a

Creator-God to whom they might be held accountable. As it was then, so it is today. As we shall see in later chapters, evolution is at its core a moral issue and not the biological panacea as it is touted. How else can one rationally account for the misinformation that continues to surround this important issue? What we see today regarding the inflated importance of both Darwin and his ideas should be expected for it was foretold in God's Word. The Apostle Paul not only foresees the present situation, but provides a rational explanation for why it is so.

For since the creation of the world God's invisible qualities-- his eternal power and divine nature-- have been clearly seen, being understood from what has been made, so that men are without excuse. For although they knew God, they neither glorified him as God nor gave thanks to him, but their thinking became futile and their foolish hearts were darkened. Although they claimed to be wise, they became fools. (Rom 1:20-22)

In spite of what students read in their biology textbooks, are told in the classrooms and what we all hear from the media, cracks are forming in the evolution foundation. Some are predicting the colossal superstructure is teetering on the brink of collapse. I strongly agree. A growing number of respected scientists are doubting materialistic evolution alone can adequately explain the origin of life or the extreme complexity of living things. This is due in large part to our increased knowledge about the intricities of living organisms especially at the subcelluar level and our explosive knowledge of genetics. Dr. Jerry Bergman and others have compiled lists of Darwin skeptics now numbering over three thousand scientists and the number is growing each month. The actual number of scientists doubting Darwin may be far more. Many prefer to remain anonymous for fear of repercussions if their doubts were made public. This fact alone speaks volumes to the lack of objectivity in modern science. Recently, the well respected and published stellar astronomer Guillermo was denied tenure at Iowa State University for writing a Christian book, the **Privileged Planet**. The risks for openly being a Christian and believing the Biblical account of Creation are increasing. Such are the times we live in.

Francisco Ayala is professor of biological sciences and of philosophy at the University of California, Irvine. He is a militant spokesperson for evolution, yet he recently stated in the prestigious **Proceeding for the National Academy of Sciences,** an apologetic for macroevolution that upon investigation, evolution contains no real science.

Darwin's greatest contribution to science is that he completed the Copernican Revolution by drawing out for biology the notion of nature as a system of matter in motion governed by natural laws. With Darwin's discovery of natural selection, the origin and adaptations of organisms were brought into the realm of science. The adaptive features of organisms could now be explained, like the phenomena of the inanimate world, as the result of natural processes, without recourse to an Intelligent Designer.

The Copernican and the Darwinian Revolutions may be seen as the two stages of the one Scientific Revolution. They jointly ushered in the beginning of science in the modern sense of the word: explanation through natural laws. Darwin's theory of natural selection accounts for the "design" of organisms, and for their wondrous diversity, as the result of natural processes, the gradual accumulation of spontaneously arisen variations (mutations) sorted out by natural selection. Which characteristics will be selected depends on which variations happen to be present at a given time in a given place. This in turn depends on the random process of mutation as well as on the previous history of the organisms. Mutation and selection have jointly driven the marvelous process that, starting from microscopic organisms, has yielded orchids, birds, and humans (Ayala, 2007).

Nevertheless, nothing in the above quote can be labeled as observable, testable empirical science. It is instead a statement of faith and shows a reverence, even worship of Saint Darwin. Still, macroevolution and its make-believe history have been thoroughly entrenched in the minds of millions as absolute fact. For decades, there were facets of this philosophy one was never to question, but irritating scientific discoveries continue to unravel the Darwinian garment. Those pesky facts continue to shake the foundation of evolution dogma. With all the new discoveries in cell biology it is not a good time to be an evolutionist.

Here is another telling example. A few decades ago, it was routinely and dogmatically taught that vertebrates arose long after the Cambrian period. Evolutionists maintained that the Cambrian, beginning 542 million years ago, was when "simple" life was first getting established. It would take many millions of years to produce the first animals with backbones, the fish. In fact, two evolutionists stated in a well-known text: *Fish arose during the Ordovician |beginning '488 mya'* (Ayala & Valentine, 1979).

However, in 1999, fossil fish were found in *lower* Cambrian sediments in south China. The following is a personal communication from Frank Sherman of the Institute for Creation Research. He attended the International Conference on Dinosaur/Bird Evolution in Ft. Lauderdale. One afternoon, a number of the conference participants took a field trip led by a recognized "expert." He asked us if the field in which we were standing could have been a dinosaur-age environment. Several said no, *because there was grass present*. Evolutionists maintain that grasses were not present during the age of dinosaurs: *In my review i.e., Eschberger, editor of Disney's new movie "Dinosaur," I mentioned that one of the few scientific inaccuracies that I found in the movie was the presence of grasses in the dinosaur nesting grounds* (Eschberger, 2000).

However, in 2005, a CBS News report said, "Plant-eating dinosaurs munched on grass, say scientists who had thought the plants emerged after the beasts died off." Evolution dogma must be continually revised to agree with newly uncovered and often damning facts.

Another example is revealing. Students were taught the only mammals during the "age of dinosaurs" were small, and barely able to stay alive among the terrible thunder lizards. Evolution theory said that the mammals were nothing more than "shrew-like insectivores that hunted at night." That radically changed with the recent discovery of large, dinosaur-hunting mammals! (Hecht, 2005) Again, the only constant in evolution seems to be change and revision. Perhaps soon there will be an apology. It is long overdue, but I will not hold my breath.

One of the more spectacular discoveries that have done much to dispel the myth of dinosaurs living many millions of years ago is the unearthing of soft dinosaur tissue (Schweitzer, 2005). How could dinosaur tissue remain soft for 70 million years?

These discoveries, while devastating Darwinism, clearly support the creation model, with all things created a few thousand years ago. Perhaps Steven Spielberg's classic dinosaur movie, *Jurassic Park* will someday come to pass. I would like that very much.

Dover

Perhaps the most important legal ruling of the first part of the twenty-first century involved the widely publicized Dover, Pennsylvania case. It is fair to say that most judges residing on state and local courts have a secular worldview. This is reflected in education decisions by the courts that consistently rule against a fair representation of creation science and intelligent design, even to the extent of not allowing public school teachers the freedom to criticize evolution. This closed-minded approach seems un-American and is unprecedented in other areas of education. It says a great deal about the lack of confidence evolutionists have for the scientific evidence supporting their worldview.

One of the latest decisions censoring a *non*-biblical origins account was the infamous *Tammy Kitzmiller, et al. v. Dover Area School District, et al.* case of December 20, 2005. U.S. District Judge John Jones issued a 139 page decision that was both mistaken and angry at many points when addressing ID. Evolutionists such as Francisco Ayala mistakenly labeled Judge Jones as an evangelical Christian, but the strident tenor of his infamous decision clearly showed this was not true. For example, at no point were the advocates of a two model approach trying to censor evolution, they were simply trying to offer a non-biblical alternative for public school students and for the Pennsylvania School Board to consider. Their plea was ignored.

In one place Judge Jones wrote, "We find that while ID arguments may be true, ID is not science" (p.64). The same can be said with even greater force and beliefs regarding macroevolution as currently taught using taxpayer dollars. Yet, it alone reigns supreme without legal challenge. Macroevolution is not empirical science. There is no experimental proof. Furthermore, throughout the ruling, Jones equated ID with creationism, an equivalence both sides strongly deny. While all creationists believe in intelligent design, not all ID proponents are creationists. Many people having a non-biblical worldview have joined the ID movement largely because it is not biblically based.

My friend and colleague Frank Sherwin, was able to be on Lou Dobbs' *CNN Live* where he debated an atheist regarding the Dover case. The point made in the five minute interview was that if ID in Dover was allegedly trying to bring 'creationism' through the back door of American public schools, then evolution, as currently taught in American public schools, is successfully bringing atheism through the front door!

Those that read Jones' decision were struck by the tired and familiar arguments that have been used in the past against creation science. Judge Jones' long ruling included a selective history of past science cases. It all came down to how one defines the term science. The modern definition of science as used by the courts and most scientists no longer includes the phrase "the search for truth," but only atheistic naturalism. However, as the outspoken and well-known militant atheist Richard Dawkins stated, "Biology is the study of complicated things that give the appearance of having been designed for a purpose" (Dawkins, 1987). Students (including those in Dover) should be allowed to understand that design implies a Designer, and something that looks created could indeed have a Creator. There should be nothing to stop a person from following wherever the scientific evidence leads even if that path leads to design. To stop a student from considering the design inference is to teach an atheistic worldview where there is nothing but molecules in motion. Our students must be taught there are thousands of reputable scientists worldwide that have rejected evolution for lack of evidence and not for religious reasons.

Faith and science were defined as a false dichotomy in Jones' opinion, being placed in the realm of either unsupported belief or of observational truth. However, both those affiliated with ID and creationists state their beliefs are factual, based on observational science. No one, neither Darwinists nor non-evolutionists, actually observed the origin of life or the origin of any of the major groups of plants or animals. Scientific observations agree with the design (or creation) argument, and not materialistic macroevolution. For example, there are clearly observed natural limits to biological change. Censoring or removal of non-Darwinian, observation-based origins models in taxpayer-paid schools is not education, but indoctrination of the atheistic model of origins. It is not freedom from religion as advocates proclaim, but is instead a form of state sponsored religion, that of human secularism based firmly on naturalistic evolution. One religious viewpoint has replaced another. The message is not freedom from religion, but freedom from Christianity deemed by the ruling intelligentsia to be the wrong religion.

Creation and ID came under attack again when a major Public Broadcasting Service NOVA documentary, *Judgment Day Intelligent Design on Trial*, broadcast nation-wide on November 13, 2007. It cast the event as a major victory for science, while Bible believing Christians and and those accepting Creation or ID were once again dismissed as ignorant and uneducated. The documentary conveniently failed to mention that the scientists supporting evolution boycotted the event and failed to show up for cross examination. Undoubtedly, they knew their position was indefensible when the facts were looked at closely. It also failed to say anything about the importance of Christian scientists in the history of science, for they accepted the Genesis account of Creation as factual.

Sadly, this video has been provided to every public school in the United States along with an extremely biased instructional teacher's manual. The reference material is introduced with a cartoon many Christians found highly objectionable claiming the views of Biblical Creation, Creationism and Intelligent Design had to be discarded in the light of evolution. The stated purpose of the supplemental material was to prepare teachers for the challenges they might encounter in teaching evolution. It is regrettable that few teachers will question the objectivity or authority of the PBS material. Fewer still have the courage and training to dispute the national educational establishment including NOVA, no matter how biased and objectionable the material.

Creation scientists continue to maintain that creation (or ID) is every bit as scientific as macroevolution, and that macroevolution is as religious as ID (or creation). When it comes to origins, both sides require faith. Advocates of ID and creationists must continue to demand their views of origins are heard in America's public schools as long as American taxpayers are required to pay the cost of education. This issue is far from dead. Many feel the tide of public opinion is beginning to rise. The recent release of Ben Stein's **Expelled, No Intelligence Allowed** has brought this issue into the public arena in a way few books or television programs could have done. Let's keep this discussion alive.

Although the Creation/ID side has lost the legal battles recently, there is an upside. Young people are often attracted to the forbidden. Banning Creation/ID from the public schools has driven untold thousands of teenagers to investigate the topic on the Internet and in libraries. Perhaps the strongest evidence of this is the explosion of IDEA (pro-creation clubs) in high school and university campuses all over the country. Recent surveys show interest in campus religious activities is currently the highest it has been in decades. May this trend continue and grow ever stronger. Truth reigns.

Is Secular Science Objective?

Students in public schools and universities are taught science is more objective than any other human endeavor. Science, they are told, deals only with facts that can be proven based on observations and experiments that can be repeated. This is simply not true. Scientists, like other humans, carry their own preconceptions and prejudices. They are influenced by their education, background and worldviews. Prejudice abounds, even in science. Modern secular scientists are free to follow the evidence only if it lends support to materialistic evolution. For example, we have an abundance of hard physical evidence for life in the past found in the massive layers of sedimentary rocks throughout the world. The billions of fossils these layers contain do not document gradual evolution, but instead shout of a rapid and complete Creation and subsequent global flood of water, exactly as mentioned in scripture. Yet scientists that interpret the evidence truthfully are openly ridiculed and many have been fired or denied tenure. Science is far from objective when it comes to evolution. In many ways, evolution has become a sacred cow in science and any attack is considered sacrilege.

Charles Darwin was a naturalist who worked in a variety of areas such as plant breeding and about eight years of research on barnacles. However, when it came to fossils (physical evidence), documenting descent with modification, or macroevolution, he knew they argued against his idea. In fact, chapter ten in his 1859 book is entitled, "On the Imperfection of the Geologic Record."

What Darwin did was profoundly unscientific. He decided to ignore the physical evidence of the fossil record and hold tightly to his unproved, unobserved theory of descent with modification known today as macroevolution. Scientists do not normally do this, but Darwin felt time and continued research would vindicate his position. The opposite actually occurred. As many scientists know, paleontological investigation through the decades has been devastating to Darwin (Swift, 2002).

The philosophy of Darwinism has prevented many from viewing clear physical evidence and making an unbiased determination based on that evidence. This was graphically displayed recently when "70 million year old" soft dinosaur tissue was unearthed in eastern Montana. Indeed, *National Geographic News* admitted many dinosaur fossils could have soft tissue inside (Norris, 2006). It is painfully obvious that organic tissue cannot maintain such pristine condition through those enormous time spans required by evolution. Something is obviously amiss. The only alternative explanation is the tissue and surrounding fossilized material is quite young and not the millions of years demanded by secular evolutionists. The philosophical convictions of these scientists will not allow them to go where the physical evidence clearly leads.

Many strongly believe a philosophical framework is not something that can be eliminated in order to provide 'objectivity.' It is clearly evident that, 'objectivity' does not exist in science when it comes to evolution. The scientist's worldview influences decisions about what experiments to conduct, which data to collect and how to interpret the results. ***Discover*** magazine ran an article regarding Dr. Mary Schweitzer's phenomenal soft dinosaur tissue discovery. She stated: *I had one reviewer tell me he didn't care what the data said; he knew that what I was finding wasn't possible. I wrote back and said, "Well, what data would convince you?" And he said, "None."* (Yeoman, 2006). There is no need to describe the secular community's outrage if a creation scientist were to say such a thing! Scientists of all stripes should have the freedom to go where the evidence leads. *As the result of the impact of what T. H. Huxley once called 'one ugly little fact,' the truly scientific investigator will make whatever changes are demanded by the situation, even if he is compelled to begin his research de novo to all intents and purposes* (Harrison, ??).

There is perhaps an even better example of the lack of objectivity in modern culture. The world was shocked on June 25, 2008 when the Spanish parliament approved a resolution extending human rights to apes. If this becomes law, it is perhaps the ultimate application of evolution to society and sets legal precedence blurring any line separating man and beast. Evolution teaches man is nothing more than an animal and such a law is a logical application of such a philosophy. Contrast this apostate mindset with what the Word of God says about man. **You made him a little lower than the heavenly beings and crowned him with glory and honor** (Ps 8:5, NIV). Man is more than a mere animal. We were created to be the image bearers of the Creator of the universe.

A goal of the intelligent design movement is to bring science back into the mainstream, where those who are not atheists may conduct research and investigation free from the bias of a purely materialistic worldview. Ideas have consequences and for too long, those who entered into scientific research or education with its roots in atheism or naturalism experienced pressure to seek only materialistic explanations. The Apostle Paul's warning to young Timothy are even more appropriate today. ***O Timothy, keep that which is committed to thy trust, avoiding profane and vain babblings, and oppositions of science falsely so called: which some professing have erred concerning the faith.*** (1 Tim 6:20-21, KJV)

What we see in society today is the natural outgrowth or consequence is a values-free, purpose-free worldview that can only be materialistic. For a majority in society that have a theistic worldview, this will not do. To insist on a purely atheistic interpretation of science is beyond the pale. Science as we know it will suffer and die if scientists do not again have the freedom to follow the evidence no matter where it leads. Again, there is a growing number of Darwin doubters and they continue to speak out presenting the case for an alternative view of Darwinism that, for many, includes Biblical creation.

The current revolution in creation, catastrophism and in the evaluation of scientific paradigms are due in no small part to Morris & Whitcomb's groundbreaking *The Genesis Flood* book. The philosophy of logical positivism has crumbled before logic and science rightly divided. People once diffident regarding origins, now clearly see the case for creation, and the Creator who has revealed Himself in scripture, nature and in the person of Jesus Christ.

Although Intelligent Design has its place in society by those who feel comfortable with it, for the Christian, it is but the first step in the stairway leading to the creation account detailed in Genesis. The eternal issue is saving faith in the One who created us in His image. This is not addressed by proponents of ID. Biblical truth (See Acts 17:11) must not be abandoned for a generic model that questions the evolutionary paradigm and leaves the door cracked for the possibility of some vague ethereal creator. Everyone can see evidence for creation/Intelligent Design (Romans 1:20), but we must boldly identify who this Creator is (Colossians 1:16; 2:3) and how one may get to know Him (Romans 10:9). It would be tragic indeed for someone to place their life's work into exposing the flaws of Darwinism and champion ID, only to die in their sins because they refused to investigate who the Designer is!

Furthermore, it seems as this new century begins, when all points of view are urged to be discussed and expressed, Christians and non-Darwinists are banned from the table of discourse. The Christian worldview is censored from today's intolerant PC culture (not to mention many churches!). But believers are warned by the Lord Himself in Luke 9:26, ***"Whosoever shall be ashamed of me and of my words, of Him shall the Son of man be ashamed, when He shall come in His own glory."*** As we speak to atheists and others with a secular worldview regarding evidence for Creation or Intelligent Design, let us remember for example, that Paul did not deviate from his unpopular message on Mars Hill, but instead, he openly presented to the unbelievers exactly who the Creator was (Acts 17:16-18). Christians in the twenty-first century must do no less. ***I am not ashamed of the gospel, because it is the power of God for the salvation of everyone who believes: first for the Jew, then for the Gentile.*** (Romans 1:16, NIV) We must follow these words from Peter. ***Always be prepared to give an answer to everyone who asks you to give the reason for the hope that you have. But do this with gentleness and respect.*** (1 Pet 3:15b)

Creator of All Things.

Chapter 9
Theistic Evolution

They exchanged the truth of God for a lie, and worshiped and served created things rather than the Creator-- who is forever praised. Amen. (Rom 1:25)

Theistic evolution has been an enigma for me since my college days. It is an unsuccessful attempt to reconcile secular science with the Biblical account of Creation. The two worldviews are opposite and agreement is impossible. I have long seen it as groundless and even dishonest. It is nothing more than a feeble attempt by some Bible scholars and Christians lacking knowledge of science to reconcile the scientifically accepted atheistic evolution with scripture by changing the time scale. It has neither the support of secular science nor of scripture. God's Word clearly states all things were created in six literal days. It was not a long drawn out process, but a miracle. The length of each the day is defined as "evening and morning" making it obvious to all the days of creation were literal twenty-four hour days. The meaning of the day length in the Creation account could not be more clear or easier to comprehend. The day length is reaffirmed again in Exodus. *Remember the Sabbath day by keeping it holy. Six days you shall labor and do all your work, but the seventh day is a Sabbath to the LORD your God. On it you shall not do any work, neither you, nor your son or daughter, nor your manservant or maidservant, nor your animals, nor the alien within your gates. For in six days the LORD made the heavens and the earth, the sea, and all that is in them, but he rested on the seventh day* (Exodus 20:8-11, NIV). Obviously, refraining from work on the Sabbath day loses all meaning if the days in Genesis were anything but literal twenty-four hour days. Any attempt to distort the six days of Creation into the vast periods of time demanded by evolutionists is without scriptural support. Nor is it compatible with evolution dogma.

Evolution is always defined as a fully natural process devoid of supernatural intervention or direction. If God, or any form of Intelligent Designer, played a role in the process then the process is simply not evolution. There is no middle ground. The source and purpose of the Bible is very clear. *All scripture is inspired by God and profitable for teaching, for reproof, for correction, for training in righteousness: that the man of God may be adequate, equipped for every good work* (2 Tim 3:16-17). Paul warned us to: *guard what has been entrusted to you, avoiding worldly and empty chatter and opposing arguments of what is falsely called "science"- which some have professed and thus gone astray from the faith.* (1 Tim 6:20-21). Finally, Paul warns us in 2 Tim 3:7 that in these: *last days difficult times will come when men will be...Always learning and never able to come to the knowledge of the truth.* This statement seems especially fitting regarding origins, for knowledge indeed abounds, yet truth remains elusive.

Let's consider some of the "empty chatter" the Bible calls modern secular science. It is important to know what the leaders of evolution think about God before we let them influence our own interpretation of scripture or our relationship with the God of Creation. Christian young people in secular universities often overlook this point in their eagerness to learn all they can from their textbooks and professors. What do leading evolutionists say about God and Creation? Given the importance of science in our daily lives this is a relevant point and will be considered at some depth.

Science has given our modern world so much that it seems few people doubt what scientists say in other matters as well. Let me share an example. Shortly after receiving my doctorate degree in zoology from Texas Tech, I was attending a Wednesday night church business meeting. A heated discussion regarding the purchase of choir robes ensued. Someone remembered my newly acquired doctorate and gave me the floor. Yes, I had recently learned some things about alligator thermoregulation, but had absolutely nothing to offer regarding choir robes, yet they thought my education gave me all knowledge. So it is with scientists. Each scientist has his or her own area of expertise, but we must not assume knowledge and training in one narrow area of science gives them wisdom in other areas of in spiritual things.

Let me state unequivocally it has been my personal experience as a biology student in dozens of undergraduate and graduate classes and as a guest lecturer in scores of universities and scientific meetings on three continents that evolution is *always* defined by evolutionists as a fully natural process. Many Christians think evolution is a scientific way of viewing Creation. Some see evolution as the method God used in Creation. Nothing could be farther from the truth! It immediately raises doubts about the time frame of the Creation week. If Creation took millions or billions of years as evolutionists claim, then our whole concept of the Sabbath is without meaning. We will return to this topic after some examples and farther development. Creation clearly was a miracle and *not* a slow gradual process. Evolution as understood and defined by the world's leading evolutionists totally excludes any intervention by an intelligent designer or the God of the Bible. Since this is an important point and central to our discussion let's consider some examples.

A leading children's encyclopedia says, ***There is probably not a scientist living today that does not believe in evolution.*** I read this while I was in fifth grade and was devastated, for I firmly believed in God and the Bible, yet wanted to become a scientist. I was lucky, for it only took fifteen years for me to recognize that statement was a lie from the Enemy and not true. Indeed, there are now tens of thousands of reputable scientists around the world that reject evolution for scientific reasons. Unfortunately, most students never recognize this teaching as a lie and many remain forever lost or their Christian growth and Bible study dwindles and eventually stops.

From the Botany textbook I had as a student: In the years following Darwin's Origin of the species (1859), the theory of evolution gradually replaced the concept of special Creation. Notice in science that evolution and Creation are mutually exclusive. It is one or the other, not both! Again, theistic evolutionists lack support from the Bible and from secular science. They stand alone, on shifting sand.

Let me illustrate. In an attempt to counter the statement that all scientists accept evolution, some Christian teacher's make statements like, "The true leaders of science see God in nature." Although it sounds good, it is simply not true. While teaching a graduate course in Invertebrate Paleontology at the Institute for Creation Research (ICR) in San Diego, California, I arranged to take my class for a behind-the-scenes tour of Scripps Institution of Oceanography at nearby La Jolla. The research director, Dr. Fred White had been my major professor while working on my Ph.D. at UCLA and we kept in touch. Dr. White was a scientist's scientist with hundreds of published papers in several different areas. All of the ICR students were from conservative Christian colleges and had a deep interest in science. Many of them were home schooled. Most of them wanted to teach science in Christian schools.

Dr. White provided an excellent tour that lasted the entire day. This is a world renowned research facility and he took the time to explain each of the research projects going on at Scripps. He also told them about some of his own reptilian thermoregulation research from UCLA and how his own work with reptiles had helped reduce infant mortality during heart valve repair by introducing the lowering of the infant's body temperature during surgery. This made the babies reptilian-like, reducing their need for oxygen, thereby greatly extending the time they could survive without blood flow. He even related an incident at the sea lion tank where the Queen of England was nearly bitten by an overly friendly sea lion.

After this once-in-a-lifetime tour, we discussed some of the things they had been shown. Of course I was hoping the students had been impressed with the level of scientific research they had just witnessed. Instead, they were most impressed that here was a leading world-class scientist who actually accepted evolution, and did so wholeheartedly. For Dr. White, like most other scientists today, evolution was the common denominator that pulled all of science and his world together. His endless illustrations and evolution charts in his laboratories and office made this abundantly clear. It was an eye opening experience for these naive Christian students. For those of you that have had similarly sheltered lives, that is what the real world of secular science is like and you must be prepared to deal with it. You must learn to do so without compromising your faith in the Word of God or your own personal relation with the Creator. Let me share one more personal experience to illustrate this point.

Upon my arrival at UCLA as a new doctorial student, I was required to take an intense qualification exam. It was oral and lasted several hours. I stood alone facing my own highly intimidating graduate committee plus several "visiting professors" who seemed to take great delight in humiliating any new student. In rapid order, I was asked a diversity of questions on all life science courses I had taken plus other areas they thought I should know in order to be permitted study at such a prestigious university. Fully 80 percent of the questions dealt with evolution because they thought it was the most important topic and fundamental to any doctorial student's success. Due to years of intense study in this area, albeit mostly looking at the "other side," I was the only incoming Ph.D. candidate that year out of over a hundred to pass the qualifying examination without being required to take remedial course work. Evolution is seen as the common denominator tying all living things together and evolution alone gives meaning to life science.

Prepare, but be wary young people! Heed the words of Christ: *I am sending you out like sheep among wolves. Therefore be as shrewd as snakes and as innocent as doves.* (Matt 10:16, NIV) Know why you believe what you believe and be able to defend it later after you have earned your terminal degree. Your role as a university student is to learn about evolution for it has influenced all of secular science. Think of it in the same way you might learn about the Greek and Roman gods. You can learn about them without worshiping them and you can learn about evolution without accepting it as a worldview. Let me also emphasize that your role as student is to learn and *not* to argue with your professors. Pray for them, but any arguments you attempt will fall on deaf ears and will only bring ridicule or worse. You must also be very careful discussing your view of origins with students. If you feel God wants you to earn a graduate degree at a secular university, go for it with all your ability and focus, but this is *not* the time to cause ripples. Keep in mind that graduate degrees are never earned in the sense an undergraduate degree is earned, by simply accumulating the necessary number of course credits and grades. Graduate degrees are bestowed at the pleasure of your graduate committee. It is highly subjective. Graduate school is not the time or the place to make enemies. As a creationist you will have ample time to attend to that after graduation. Earn that degree so your voice will be heard.

I personally know an education doctorial student at Oklahoma University who was denied her degree because she would not lead her junior high class to search for their own personal spirit guide (Read: "Demon Possession"). Nor would I have done it, but not every issue is worth going to the ditch for. Do not look for trouble…or see your secular university as a mission field until you have that degree on the wall.

Remember my own personal experience and learn from it. Again, two months after I graduated with a master's degree in biology from Baylor University, my graduate committee found out I was a Creationist and my graduate committee chairman phoned me to inform me in no uncertain terms that they would never have accepted me for graduate study at Baylor had they known I did not accept evolution. Two weeks after I graduated with my Ph.D. in zoology from Texas Tech, I had another article published in the ***Creation Research Society Quarterly*** and my major professor found out. He actually formed a committee to annul my doctorate, but was unsuccessful. Our battle is real.

Remember the warning of the Apostle Paul, ***In fact, everyone who wants to live a godly life in Christ Jesus will be persecuted.*** (2 Tim 3:12) As a result of these two incidents, I was not able to get a letter of recommendation from either of these graduate schools, nor do I feel welcome to return to either school. Be forewarned and seek Divine wisdom and supernatural protection. It was only by the grace of God that I survived both attacks.

Two things should now be obvious to those of you who have endured to this point. First, evolution as defined by leading evolutionists is a fully natural process. There is no room for a First Cause. Secondly, the reason evolution was and is so eagerly embraced is not the scientific evidence, but instead, is at the core a moral issue.

Four World Views

There are four major positions regarding the origin of the universe and the origin and diversity of life: atheistic evolution, theistic evolution, intelligent design and supernatural Creation. There are a variety of intergrades or modifications such as the gap or the day-age theory.

Atheistic evolution was discussed in chapters seven and eight and is the belief that matter and natural law is eternal and sufficient to account for the complexities found in the universe and for the origin and diversity of living things including man. There is no need or room for God and thus one can do as he or she pleases for there is no absolute truth, no sin, no life after death and no judgment. This is what the world's leading evolutionists and most scientists' embrace. This is what most secular university professors and textbook authors accept. This is the teaching that permeates modern science from biology to physics to history and the social sciences. This is what the Living God warned us about repeatedly. It is indeed the majority view held by most scientists in the world today.

Theistic Evolution is discussed in this chapter and is an attempt to embrace and even to reconcile the Word of God to modern secular science. Even the term "theistic evolution" is a contradiction, an oxymoron for "theistic" means "belief in the existence of God view as the creative source of man and the world who transcends yet is imminent in the world" (From the Merrian-Webster's Collegiate Dictionary) while evolution is always defined as a natural process apart from God. Conversely, if Creation was not a miracle, if it can be explained by natural law it is *not* Creation. Or seen another way, the theistic evolutionist has neither the support of leading evolutionary scientists nor that of conservative Bible scholars. It is but an impotent attempt to reconcile one of conflicts between modern science and the Bible. Even if you vehemently disagree with this statement, please stay tuned and let me show you how weak are the evidences for evolution for it is surely a house built on shifting sand. Many in the current ID movement feel that at some future time scientists will look back and wonder how such a technologically advanced society could be so deceived into accepting such a scientifically unsubstantiated claim as evolution. We agree wholeheartedly. The Apostle Peter prophesied its acceptance in these end times and said the reason would be *"Willful ignorance"* (II Peter 3:3-7).

Intelligent Design will be discussed in chapter eleven. It is the newest viewpoint and is growing rapidly. It is not Bible based, but is based on recent scientific discoveries that living organisms, especially at the subcellular and molecular level is far to complex to have arisen by random events. Such complexity demands a designer, but that designer is not defined or identified.

Supernatural Creation is the major content in this book and is plainly taught in scripture. It was generally accepted by the scientific community prior to Charles Darwin, but is openly ridiculed today by many in academia. Nevertheless, thousands of reputable scientists around the world embrace it. Unfortunately, the risks of doing so are high and over three thousand of us have been denied tenure and are black listed by the ACLU and others.

Let me close this somewhat dismal discussion with a deeply profound poem. It seems people in the arts are often more accurate and forthright in assessing contemporary society than are academicians. The following insightful satirical poem on the modern mindset was written by the British journalist Steve Turner and is quoted in a recent book by Ravi Zacharias (Ravi Zacharias International Ministries). The poem has become acutely relevant in the midst of our current epidemic of school violence and rash of drive by shootings.

CREED

We believe in Marxfreudanddarwin.
We believe everything is OK
as long as you don't hurt anyone,
to the best of your definition of hurt,
and to the best of your knowledge.

We believe in sex before, during, and after marriage.
We believe in the therapy of sin.
We believe that adultery is fun.
We believe that sodomy's OK.
We believe that taboos are taboo.

We believe everything's getting better
despite evidence to the contrary.
The evidence must be investigated
And you can prove anything with evidence.

We believe there's something in horoscopes,
UFO's and bent spoons;
Jesus was a good man just like Buddha,
Mohammed, and ourselves.
He was a good moral teacher although we think
His good morals were bad.

We believe that all religions are basically the same--
at least the one that we read was.
They all believe in love and goodness,
They only differ on matters of creation,
sin, heaven, hell, God and salvation.

We believe that after death comes Nothing
Because when you ask the dead what happens
they say nothing.
If death is not the end, if the dead have lied, then it's
compulsory heaven for all
excepting perhaps
Hitler, Stalin, and Genghis Khan.

We believe in Masters and Johnson.
What's selected is average.

239

What's average is normal.
What's normal is good.

We believe that man is essentially good.
It is only his behavior that lets him down.
This is the fault of society.
Society is the fault of conditions.
Conditions are the fault of society.

We believe that each man must find the truth that
is right for him.
Reality will adapt accordingly.
The universe will readjust.
History will alter.
We believe that there is no absolute truth
excepting the truth
that there is no absolute truth.

We believe in the rejection of creeds,
and the flowering of individual thought.

If Chance be the Father of all flesh
disaster is His rainbow in the sky,
and when you hear
State of Emergency!
Sniper Kills Ten!
Troops on Rampage!
Whites go Looting!
Bomb Blasts School!
It is but the sound of Man
Worshipping his Maker.

Chapter 10
Evolution As Idolatry

You shall have no other gods before me...for I, the LORD your God, am a jealous God. (Exod 20:3, 5)

As a zoologist, I love nature, but it sickens me to watch the nature programs on PBS educational television. The photography is speculator and I love seeing and learning about God's Creation, but educational television almost always deifies nature and glorifies natural law and the creature rather than the Creator. One often hears statements such as; natural selection is "clever," "creative" or "inventive." Nature and mutation are given credit for the beauty and diversity found in the living world, credit that should go to the God of Creation. The laws of nature are instead touted as the Master Designer. They ignore the God that created those natural laws. The Apostle Paul warned of this very thing. *They exchanged the truth of God for a lie, and worshiped and served created things rather than the Creator-- who is forever praised. Amen.* (Rom 1:25)

I have long seen the acceptance of evolution as the explanation for the origin of life and complexity of living things as a modern form of idolatry. For some this may seem to be a stretch. Let me share my reasoning. Instead of bowing down and worshiping some graven image, many today worship natural law, science or simply nature. The term "Mother Earth" is used with reverence and to the New Agers it has become a false religion. With the teaching of evolution, the creature (natural law) is praised and worshiped instead of the Creator. God created the physical laws of the universe as well as every living plant and animal, including man by the power of His spoken word. Neither they, nor we evolved. Creation was a miracle, not a slow natural process.

We should not be surprised for the apostle Peter not only predicts the acceptance of evolution in these last days, but gives the reason for its widespread acceptance. *Knowing this first, that there shall come in the last days scoffers walking after their own lusts, and saying, Where is the promise of his coming? For since the fathers fell asleep, all things continue as they were from the creation." For this they willingly are ignorant of that by the Word of God, the heavens were of old, and the earth standing out of the water and in the water by which the world that then was, being overflowed with water, perished.* (2 Pet 3:3-6, KJV) This *"willful ignorance"* dominates societal thinking today regarding Creation and its Creator. This is the best explanation for why sinful man fails to see the obvious evidence for the supernatural Creation that abounds around us. As God's inerrant Word clearly says: *The heavens declare the glory of God; the skies proclaim the work of his hands. Day after day they pour forth speech; night after night they display knowledge. There is no speech or language where their voice is not heard.* (Ps 19:1-3)   Evolutionists and society are without excuse for the glory of God are clearly seen and heard by all.

As pastor and author, Dr. Steve Kern points out in his powerful book, *No Other Gods,* that to see materialistic evolution as the creator of life is to worship natural law instead of the Living God (Kern, 2007). Science has become a god for many in our culture today. Dr. Kern goes a step further. He feels, and I certainly agree, that to reject the literal six day Creation is also a form of idolatry. Many today try to pervert the clear teaching of the Bible regarding the six day Creation and six thousand-year history of earth and man.   They have placed science over scripture. The same is true regarding many modern Bible "scholars" who assume science is all-powerful and cannot err. They attempt to read the clear account of Creation in Genesis as allegorical or poetic in order to not offend the secular science community. Not only does the original language fail to support this outlandish idea, but Jesus Christ Himself accepted a literal Creation as well as the genealogies found in Genesis. *But at the beginning of creation God 'made them male and female.'* (Mark 10:6) See also Luke 3:23-38. If we can trust Jesus regarding the more weighty matters of sin, salvation, heaven and hell, we must certainly accept His confirmation regarding the historical account of Creation and genealogies written in Genesis. Again, to do otherwise is to replace our Savior and the Living God of the Bible with the false gods of science and liberal scriptural interpretation. Our God is a jealous God. On this, the first commandment is clear to all. *You shall have no other gods before me. You shall not make for yourself an idol in the form of anything in heaven above or on the earth beneath or in the waters below. You shall not bow down to them or worship them; for I, the LORD your God, am a jealous God, punishing the children for the sin of the fathers to the third and fourth generation of those who hate me.* (Exod 20:3-5)

Sadly, many today bow down to the false god of secular science. In doing so, they risk the wrath of an angry and jealous God. The God of the Bible has shown favor to our great country for over two hundred years. Do we really want to risk putting His blessings and protection in jeopardy? Many Americans are deeply troubled by the removal of prayer and the Ten Commandments from our public schools.   In their place, our children are now forced to recite Muslim prayers. Praying to the Living God is strictly forbidden. A military chaplain was prosecuted for praying in the name of Jesus. Political correctness has gone too far. Something must be done to return to our roots…to the God of our forefathers and to the Creator of all things.

Creature Worship Today

Although the accusation of mixing Christianity with science is regularly leveled at creation scientists, one finds plenty of emotion, even religious fervor in the secular scientific community. Ornithologists in particular and biologists in general (www.cnn.com/2005/TECH/science/04/28/woodpecker/) were overjoyed recently to discover a woodpecker that was thought to have been extinct for over fifty years.

It was found alive and reproducing in the Big Woods region of Arkansas. The ivory-billed woodpecker was considered extinct until eight sightings and a videotape of the creature were compiled from 2004 to 2005. Needless to say, this caused quite a stir.

Some biologists were quite emotional in reporting the discovery. Regarding trekking in the Arkansas woods, audio archivist Martha Fischer said, "The place really is like being in a cathedral." One report said many who searched the woods and found the rare creature were forever "changed by their experience." Indeed, an associate professor of biology "put his face in his hands and began to sob" and another "was choked with emotion" after sighting the elusive bird, saying it rose "Lazarus-like from the grave."

Rediscovered plants receive the same reverent accolades. Denis Kevans and Sonia Bennett wrote a poem about a conifer thought to be extinct since Jurassic times, but was found alive in Australia in 1994.

There's a tree that's so rare,
Grows deep in the gorges out there,
Deep in my heart I will sing of the Wollemi Pine,
No preaching words, no angry tones,
The Wollemi stands all alone,
One hundred million years of passing time.

Man is inherently religious and must worship something during his sojourn on earth. If he rejects the Creator, he will likely worship the creation and those creatures that inhabit it. In popular culture and even on the occasional news report, we hear references to "Mother Earth" as creator-god. Such a view is no longer held exclusively by the tree-hugging fringe. It is becoming commonplace in mainstream America. Again, the apostle Paul warned us of this very thing. *They exchanged the truth of God for a lie, and worshiped and served created things rather than the Creator-- who is forever praised. Amen.* (Rom 1:25) Truly, this is the world we live in today for science, evolution and nature have become false gods to many.

Creation scientists certainly share in the joy of finding an animal or plant alive that was thought to be forever gone. By studying the creatures God created, we see the Hand of the Creator. We remember such things are *from the living God, who made heaven and earth and sea and everything in them.* (Acts 14:15b). He alone deserves worship and praise for the beauty of nature. From Genesis we find: *Then God blessed them, and God said to them, "Be fruitful and multiply; fill the earth and subdue it; have dominion over the fish of the sea, over the birds of the air, and over every living thing that moves on the earth.* (Gen 1:28, NKJ) Christians have a clear mandate to be wise stewards of what God has given us. Yes, God gave us dominion over the His created world from the very beginning of human existence. *Then God said, "Let us make man in our image, in our likeness, and let them rule over the fish of the sea and the birds of the air, over the livestock, over all the earth, and over all the creatures that move along the ground." So God created man in his own image, in the image of God he created him; male and female he created them. God blessed them and said to them, "Be fruitful and increase in number; fill the earth and subdue it. Rule over the fish of the sea and the birds of the air and over every living creature that moves on the ground."* (Gen 1:26-28) We have dominion over the world and are to rule over all living things, yet we must do so responsibly. We are to worship it, call it Mother, or treat it as though it is a living thing to be revered. We were commanded not to worship the creature, but to bow down and worship the One who created *every living thing that moves on the earth.* It is time we return to this fundamental and important Biblical teaching.

Chapter 11
Intelligent Design

For the foolishness of God is wiser than man's wisdom, and the weakness of God is stronger than man's strength. (1 Cor 1:25)

Intelligent Design (ID) is a relative newcomer to the origins debate. To many, it is less offensive than Bible based Creationism and has a wider acceptance than the older more fundamentalist worldview could achieve. Its proponents suggest empirical or observational information about the universe and living things provides evidence of intelligent design which better accounts for origins that does undirected evolution. Opponents of ID see it merely as new packaging for the old creationism. The modern ID movement began about twenty years after the 1961 publication of *The Genesis Flood* and is rapidly gaining popularity. It is more scientific and less theological than conventional creationism or progressive creation.

Although many holding this view are Christians, it is not based on scripture or other religious writings. The intelligent designer is not identified and therefore non-believers find it more palatable. Some of the supporters of ID are atheists. It was spawned by increasing evidence that evolution is bankrupt in accounting for a plausible mechanism for the origin of life and complexities found in living things. Intelligent Design advocates feel naturalistic evolution can be defeated on rational scientific reasons alone. Cosmologist Fred Hoyle was the first to use the term "intelligent design" in the modern sense in 1982. He said that unless a person was "deflected by a fear of incurring the wrath of scientific opinion, one arrives at the conclusion that biomaterials with their amazing measure of order must be the outcome of intelligent design" (Hoyle, 1982).

The best known and most effective advocate for ID is the Discovery Institute in Seattle, Washington (www.Discovery.org). The popular movie, *Expelled, No Intelligence Allowed* introduced the public to Discovery Institute in a positive way and interviewed some of the staff. It is a nonprofit public policy organization addressing a variety of social, political, scientific and economic issues. It was founded in 1990 by Bruce Chapman, formerly the Secretary of State for Washington and former Director of the U. S. Census Bureau under President Ronald Reagan. He was also the U. S. Ambassador to the United Nations in Vienna (See Dembski, 1998). In 1996, the Discovery Institute sponsored the organization of the Center for Science and Culture led by geologist and philosopher of science Stephen Meyer and political scientist John West.

Another important ID advocate is lawyer Philip Johnson, author of the highly successful *Darwin on Trial* and architect of the "Teach the Controversy" campaign.

Evolutionists insist the Discovery Institute is a Christian organization, but this is patently false. Discovery Institute is composed of New Agers, agnostics and Christians. One prominent member of the Institute with an earned doctorate is a lay preacher of the non-Biblical Unification Church. To be sure, the policy, writing and lectures of the Discovery Institute are strictly secular, but nonetheless are often used by atheists to make a point. Furthermore, the Discovery Institute has clearly shown secularist's strict reliance on natural causes is tantamount to the religion of naturalism. The Discovery Institute is a powerful force to reckon with and has been at the forefront in recent court debates including Dover.

Several noteworthy books have exploded on the scene advocating the virtues of ID and adding depth and scientific credibility. *The Mystery of Life's Origin* (Thaxton et. al, 1984) sharply criticized chemical evolution as inadequate in explaining the origin of living cells from non-living material. Molecular biologist Michael Denton published *Evolution: A Theory in Crisis* (Denton, 1985) the following year. In it, he criticized the evidence supporting Darwin's theory and defended the view that design could be inferred from living things. He admitted his conclusions might have religious implications, but they do not depend on religious presuppositions. *Of Pandas and People,* by biologists Percival Davis and Dean Kenyon, was edited by Charles Thaxton (Davis and Kenyon, 1991) and presented a favorable case for intelligent design. The second edition came out two years later (Davis et. al, 1993) and has continued to be widely read and glean rave reviews. Also in 1991, Berkeley law professor Phillip Johnson published *Darwin On Trial* (Johnson, 1991) which critically analyzed the logic and assumptions evolutionists use to rule out design in living things. Attorneys, with their training in critical thinking and skills in debate, seem well suited for examining the evidence for and against evolution.

Other important books supporting ID followed quickly. The ever popular *Darwin's Black Box: The Biochemical Challenge to Evolution,* published in 1996 by biochemist Michael Behe, has been particularly popular and influential. Behe argued certain aspects of the structure of living cells are characterized by "irreducible complexity" that cannot be accounted for by materialistic evolution. Certain complex features possess many parts, none of which offer any advantages or survival value alone and would have been eliminated by natural selection. He used examples of the light-sensing portion of vertebrate eyes, human blood clotting system and especially the beautifully complex bacterial flagellum. None of these examples were known to Charles Darwin, but present insurmountable obstacles for evolutionists today. No doubt, a large part of the book's success has been due to Dr. Behe's position as professor of biology at Lehigh University and his many and well attended public speaking appearances. His second book, *The Edge of Evolution: The Search for the Limits of Darwinism,* published in 2007, is also excellent and has wide appeal. Michael Behe, Phillip Johnson and other notables in the current ID movement recently co-authored, *Intelligent Design 101: Leading Experts Explain the Key Issues* and, as the title suggests, is an introduction to the broad field of ID (Behe, et. al., 2008).

If today's ID movement has a leader, it is William Dembski. As a well known philosopher/mathematician and important theorist, he is a powerful force. Dr. Dembski is truly one of the great intellectual athletes of this generation. He has a Ph.D. in math from Chicago, another Ph.D. in philosophy, plus graduate and post doctorate degrees in theology, computers and biology from Princeton, MIT and other prestigious universities. He is a true renaissance man skilled in many areas and well able to show the connections between various disciplines. He is also an excellent and entertaining author and has written many books related to ID and other topics. The timing of his writings has also been appropriate, as some in the science community have dismissed the early ID movement as "creationism in a cheap tuxedo." Today, the movement is appealing to many highly trained and deep thinking scientists and philosophers. Who better to lead these than Bill Dembski? Perhaps his best known argument is his explanatory filter. The filter utilizes a logical algorithm composed of three junction points: specificity, contingency and complexity. Intelligent Design proponents use this explanatory filter to distinguish between natural events and intelligent causation. The results overwhelmingly indicate evidence of design in living things. Several of his better known books are listed in the references section at the end of this book.

He was employed at Baylor University from 1999 to 2005 and was fired for his Intelligent Design views. The news received national attention and was seen by many as totally unfair, especially from an allegedly "Christian University." I saw it as simply business as usual for Baylor, for as I mentioned earlier I was informed I would never been allowed to attend Baylor had they know I accept the Biblical account of Creation. In June of 2006, Dr. Dembski became research professor in philosophy at Southwestern Baptist Theological Seminary in Fort Worth, Texas. I strongly encourage readers interested in learning more about the ID movement read some of Dr. Dembski's excellent books on the subject.

Important Films

In addition to many books and recent media attention, ID is also well represented in some excellent movies. *Unlocking the Mystery of Life: The Scientific Case for Intelligent Design* produced by Illustra Media (Allen, 2002) is excellent and informative. *The Privileged Planet: How our place in the cosmos is designed for discovery* (Gonzalez and Richards, 2004) has been widely seen and provides powerful arguments from astronomy. It is noteworthy that Dr. Guillermo Gonzalez was denied tenure at Iowa State University for his work on the book that led to this remarkable movie. This occurred in spite of his stellar astronomical publication record in respected peer reviewed journals. Once again, the only conclusion one can draw from his untimely dismissal is science is no longer objective and researchers do not have the freedom to follow the evidence if it leads to design or creation. Or put simply if you want to remain in academia, don't dis Darwin.

Although less widely known and more openly Christian, Louie Giglio has produced an outstanding DVD that continues where **Privileged Planet** stops. **How Great is our God** provides an excellent series of demonstrations of how really large some of the stars are when compared to earth and our own sun. A second even more powerful DVD is simply entitled **Indescribable**. The series is available from www.268generation.com and it is strongly recommended.

Perhaps the best known movie dealing with ID was Ben Stein's **Expelled, No Intelligence Allowed** released in May of 2008. This was a very important movie and brought to public awareness the lack of objectivity in science few knew was so commonplace in science today. It also showed some of the terrible influences evolution has had over the years including the German death camps and slaughter of 5 million Jews by Adolf Hitler in order to "improve the species." Perhaps more importantly, this movie has opened public discussion on the whole topic of origins and has raised serious questions about the nearly universal blind acceptance of evolution dogma by the scientific community. It is certainly worth owning and sharing.

The idea of Intelligent Design is certainly not new. Since the times of Roman and Greek philosophers Plato, Socrates and others, nature has been interpreted into two conflicting worldviews: supernatural Creation or Intelligent Design and some form of materialistic evolution now known as neo-Darwinism. William Paley (1743–1805) presented ID of his day with the publication of **Natural Theology** (1802) giving the popular and persuasive example of a watch requiring a watchmaker. However, in 1859 Darwin robbed God of worship as Creator by suggesting the more cryptic process of natural selection and natural law could produce the complex features seen in living systems. To the relief of many, the Creator was no longer needed. To many, this had great appeal and caught on quickly. Everything in the living world is now assumed by evolutionists to be the result of the accumulation of small changes over vast periods of time. God has been removed from the origins equation. Prominent atheists such as Richard Dawkins confidently state, **Biology is the study of complicated things that give the appearance of having been designed for a purpose** (Dawkins, 1987). Nevertheless, there are a growing number of scientists disenchanted with the continuing lack of actual scientific evidence supporting the evolutionary origin of life and complexity of biological systems. They consider design a rational and more intellectually satisfying alternative to Darwin's unproven and unsupported theory.

Intelligent Design has wide appeal in part because it is a clear alternative to macroevolution, but does not identify *who* the Designer is, or *how* the intelligent design came about. Nor does this movement directly address the age of the Earth, an issue many prefer to ignore. For these and other reasons, many Christians felt excluded from ID circles, a misunderstanding that was built upon and magnified by the secular community. For example, secularists attempt to drive a wedge in the ID/creation camp by saying *all* ID proponents believe in an old earth. This is offensive to some and blatantly incorrect.

It must be understood, however, that even the most orthodox Christian believes in intelligent design, produced by the master Designer Himself, the Lord Jesus Christ! Certainly scripture supports this view. ***In the beginning was the Word, and the Word was with God, and the Word was God. He was with God in the beginning. Through him all things were made; without him nothing was made that has been made.*** (John 1:1-3, NIV) As Christians battle the philosophy of evolutionism, many are happy to use the cogent arguments of the ID movement. Intelligent Design is certainly more scientific today than it was in Paley's day. Clearly, there is excellent logic in the ID paradigm. Even the secular scientific community is actively searching for evidence of intelligence beyond our solar system.

Project SETI (Search for Extra-Terrestrial Intelligence) seeks distant radio signal signatures as evidence of otherworldly intelligence. The birth of SETI came in 1955 with an article in the March issue of ***Scientific American,*** describing the concept of scanning the universe in search for extraterrestrial radio signals. Within two years, the concept was approved for construction by Ohio State University, funded by a $71,000 National Science Foundation grant. The radio telescope finally became functional in 1963 and thus began the world's first SETI program. In 1971, NASA agreed to fund an expanded SETI study known as ***Project Cyclops.*** It was to consist of a global array of 1,500 huge radio antenna dishes and came with a price tag of ten billion dollars. The government funding was delayed and eventually blocked so the project was not built, but the report formed the basis for much of the SETI work that followed. Although initial funding for SETI did come from the government, it is now largely supported by private donations. Many SETI advocates believe proof of extra-terrestrial intelligence will finally prove the Biblical account of creation wrong and destroy all remaining belief in the Christian's Creator-God.

Many questions relating to ID remain unanswered, but its influence is growing and it is well grounded in science. It is interesting that the disciplines of archaeology and forensics also have a firm foundation in ID. For example, what would an archaeologist conclude upon finding a buried pottery shard? The only logical explanation is the pottery was made by some intelligent being. The search for hard evidence continues to lend support for Intelligent Design.

Even Albert Einstein gave an ID analogy. I'm not an atheist, and I don't think I can call myself a pantheist. We are in the position of a little child entering a huge library filled with books in many languages. The child knows someone must have written those books. It does not know how. It does not understand the languages in which they are written. The child dimly suspects a mysterious order in the arrangement of the books but doesn't know what it is. That, it seems to me, is the attitude of even the most intelligent human being toward God. We see the universe marvelously arranged and obeying certain laws but only dimly understand these laws. Our limited minds grasp the mysterious force that moves the constellations. (Jammer, 2002)http://www.deism.com/Einstein1.htm)

Certainly, the assertion that the universe and all living things are the product of an Intelligent Designer is essential for any serious origins study. Such an assumption is however only the first step on the ladder that leads to the full recognition of the God of Creation. ID falls short for it fails to identify or honor the Living God of Creation and of the Bible. More importantly, ID does not lead to Jesus as Savior. It is a good start, but the journey has just begun. If we leave Christ out of the discussion, how can we truly help people who are walking in the darkness inherent in materialistic evolutionism? In their attempt to avoid criticism, Hugh Ross and others in the contemporary ID movement have so watered down the Creation, that the message and intent of the Genesis account of the Creation and Flood are meaningless. They ignore not only Genesis, but the sixty-five books that follow. Without original sin and the fall, there is no need for the Savior. What could be more important?

God's people around the world must make an important decision concerning the origin of the world. Either we take God at His Word because of who He is and knowing God never lies; or we can surrender our minds to the ever-changing opinions of finite and sinful men. Again this too was prophesized thousands of years ago. *Where is the promise of His coming? For since the fathers fell asleep, all things continue as they were from the beginning of creation.* But God gives us His infinite and eternal perspective on such thinking: *This they willingly are ignorant of…the world that then was, being overflowed with water, perished* (II Peter 3:5-6).

In the twenty-first century, we find creation science making significant inroads both within and outside the church. There are a number of reasons for this resurgence. As Christians, we see this primarily as God's hand of blessing upon the efforts of His people as many are exposing the myth of evolutionism while championing the historical account of Genesis. However, there is also another reason: *good scientific research*, whether by creationists *or* evolutionists. If it is true that all science is creation science, then as expected even secular researchers are making discoveries that reveal the case for creation. This is vehemently denied, of course, and the ever-popular secular explanation of macroevolution is offered repeatedly. Sadly, many competent scientists no longer have the freedom to follow the chain of evidence when it leads to design or Creation. If that evidence leads to recognition of God as Designer, there can lose their career as thousands have witnessed.

Nevertheless, secular scientists have made phenomenal discoveries involving sub-cellular anatomy and the processes involved. Indeed, the "poster child" of Intelligent Design and creation science is the incredibly complex rotary flagella with its numerous protein subunits looking very much like a submicroscopic axle of an eighteen-wheeler! Like so many anatomical structures large and small, it must be complete and functional to provide advantage. Ninety-five percent of the structure will not work and would be eliminated quickly by natural selection. Logic and reason fail in their claim that this ultra-tiny powered structure is simply the result of chance, natural processes and time.   God's creation is clearly seen to those who will be honest enough and having the courage to actually follow where the evidence leads. *For since the creation of the world God's invisible qualities-- his eternal power and divine nature-- have been clearly seen, being understood from what has been made, so that men are without excuse* (Rom 1:20, NIV). Leaving the submicroscopic world for the universe, we again see how scientific discoveries mirrored in the words of King David. *The heavens declare the glory of God; the skies proclaim the work of his hands. Day after day they pour forth speech; night after night they display knowledge. There is no speech or language where their voice is not heard.* (Ps 19:1-3, NIV)

Contrary to what many people today think, no one has *ever* found the Biblical account of creation to be in conflict with empirical (experimental) scientific observations. If there was a conflict, then the large number of Bible-believing creation scientists of the past would have had a problem and would have voiced it long ago. The Genesis account of creation and macroevolution do have this in common: both are beyond scientific investigation. Both must be taken by faith because they were one-time events that occurred in the unobserved past. Those events cannot be duplicated in the laboratory. Many today do not see this as a problem for we accept the truth of the Apostle Paul's message to young Timothy, *Yet I am not ashamed, because I know whom I have believed, and am convinced that he is able to guard what I have entrusted to him for that day.* (2 Tim 1:12b, NIV)

Secular Science and Intelligent Design

Intelligent Design is very unpopular with the secular segment of society. Open hostility abounds in the form of bad press, legal action and a plethora of misinformation. A double standard is seen when those holding to minority viewpoints demand recognition by lobbying their policymakers, but are outraged when those 'on the other side' attempt the very same thing. The Discovery Institute and other organizations have followed suit, campaigning to encourage public awareness of the many drawbacks of Darwinian 'science' in the press and especially in our taxpayer funded public schools.

The typical argument by evolutionists is that ID is nothing more than a wolf ("creationism") in sheep's clothing. If it is true that presenting ID in America's classroom is "creationism through the back door," then it is doubly true that the manner in which evolution is taught is bringing "atheism through the front door" of American public schools. Indeed, atheists in America are delighted and promote how science, and biology in particular, are currently taught from their narrow Darwin-based worldview.

It is obvious, however, that there are serious problems with Darwinian theory as documented in many reputable books. Some of the many scientific difficulties are documented in this book and the controversy should, and must, be taught in public schools as well. Secularists are aghast at such a suggestion, with some even going so far as to say that there are no problems with macroevolution! Just as creation scientists see no empirical validation of macroevolution, so the secular community maintains ID is not science. According to the U.S. National Academy of Sciences: *Intelligent design, and other claims of supernatural intervention in the origin of life, are not science because they cannot be tested by experiment, do not generate any predictions, and propose no new hypotheses of their own* (National Academy of Sciences, 1999). This statement may be said with equal force regarding the philosophy of Darwinism! If the truth be told, one is just as religious, or scientific, as the other. Both must be accepted largely by faith.

Intelligent Design has performed a remarkable service by showing the public that things appear to be designed and neo-Darwinism utterly fails to provide a step-by-step explanation. However, ID does not address or answer life's big questions: "Where did I come from?", "What is my purpose here on Earth?" and "Where am I going?" ID says nothing about heaven, hell or the final judgment. Some creationists are excluded by those active in the ID because of their refusal to address the age of the Earth, who the Designer is and many more of life's important questions.

Perhaps the main issue CRS, ICR and other scientific creation organizations face when interacting with the public is the matter of secular science being at risk of being thought of as only religious by any mention of

Design or Creation. What people fail to see is the religion of secular humanism has as its foundation macroevolution (the 'particles-to-people' idea) is already well established in science, is dogmatically presented in America's educational system. They also appear to be blind to the historical fact that science flourished and men of science who were also men of God made major discoveries in the past.

Chapter 12
Influence of Evolution On Society

Watch out for false prophets. They come to you in sheep's clothing, but inwardly they are ferocious wolves. By their fruit, you will recognize them.   Do people pick grapes from thornbushes, or figs from thistles? Likewise, every good tree bears good fruit, but a bad tree bears bad fruit. (Matt 7:15-17, NIV)

Ideas have consequences. This is certainly true regarding the societal impact of Charles Darwin's theory of evolution by natural selection. The world will never be the same. Let us consider some of the horrific applications of Darwin's idea of evolution. Many will be shocked because until recently, the dark truth about godless evolution was hidden from the public. The murderous application of evolution is not mentioned in textbooks or discussed in university classrooms. The media will not cover it. The public is largely unaware of the terrible consequences of the application of Darwinian evolution on society in the United States and abroad. Yet, we must learn these terrible lessons from the past if we are to make wise decisions in the future. The historical influence of evolution has been widespread and violent, yet few know of its tainted aftermath. Let us consider the deadly legacy left by Charles Darwin and contemplate carefully some of the fruits of his teaching. The scripture passage at the beginning of the chapter is particularly insightful. Let's look carefully at the fruits of evolution.

Charles Darwin (1809-1882) has achieved saint-like status throughout much of the science world today. He is considered by many to be the greatest biologist to have ever lived and was buried in the prestigious Westminster Abbey near Sir Isaac Newton. Burial in the Abbey is normally reserved for English monarchs and a few others held in high esteem. February 12th is celebrated internationally as "Darwin Day."   On Darwin Day 2009, Charles Darwin's memory was hailed in a global celebration of science and humanity never before seen. (See: www.darwinday.org)   There was even talk of making Darwin's birthday a national holiday…like President's Day. It was Darwin's 200th birthday and the 150th anniversary of the publication of his famous book, ***On the Origin of Species by Means of Natural Selection, or the Preservation of Favoured Races in the Struggle for Life.*** The book was published on November 24, 1859 and all

1,250 copies sold out within 24 hours. Let's look carefully at some of the facts and see if such profound adoration is deserved. What are some of the consequences of his idea of the origin of species by natural selection? Many will be surprised at what follows, for the dark truth has been carefully hidden from the public for decades.

In public schools and university classrooms around the world, evolution is taught as the major uniting foundation for all of modern biology. Darwin's influence reaches far outside biology into such areas as psychology, education, sociology, politics, law and even the practice of medicine and religion. I find it telling that individuals or the media can openly ridicule the President of the United States, late night comics can joke about the Pope and even deride Jesus Christ, but uncomplimentary statements about Charles Darwin or evolution are unacceptable. Tolerance is preached today, but is not practiced. As we will see in a later chapter, over a three thousand highly educated and qualified university professors have been fired for the unthinkable--for doubting Darwin's evolution. I know, for I was one of them. Even more tragic, over two hundred exceptionally bright and highly motivated graduate students have been denied access to higher education for doubting Darwin. Neither science nor education is objective.

Religious persecution exits today right here in the United States, yet we are taught this is the "Land of the Free" where free speech is protected. It is not true regarding Darwin or evolution. Many Christians today see the origins question as peripheral, an unimportant side issue. Christians *must* be informed. The application of evolution has been devastating around the world resulting in the death of untold millions. Its death and destruction continues today in the twenty-first century. Clearly, materialistic macroevolution has spawned a number of practices and some have been detrimental to society worldwide.

One of the fruits of the acceptance of evolution dogma is the unscriptural idea that man is nothing more than an animal, having evolved from lower forms of life over vast eons of time through "descent with modification" as Darwin said. As an animal, life after death is impossible and considered a myth of the uneducated. Tragically, this is the only model presented in public education and on nature shows seen on television. The secular community warns unceasingly of "ignorant creationist know-nothings." Let's consider carefully what evolution offers in place of a Creation worldview. Several diverse areas will be considered in depth.

Evolution's Influence On Students

As a life-long teacher, I have a heart for students of all ages. Removal of the Bible, prayer and the Ten Commandments from our public schools has had tragic and far-reaching effects. Today, children are taught man is the product of mindless evolution and is but an animal. Life after death is impossible and we are

256

no longer responsible for our actions. Eternal judgment, heaven and hell are considered myths of the unlearned. Life is nothing more than "matter in motion." Morality and ethics can no longer be taught in our public schools. Instead, students are taught there is no such thing as right and wrong, for everything is relative. It all depends on the person and situation. With this sort of philosophy being proclaimed to our children, we should not be surprised that school and university shootings are becoming pandemic. Metal detectors and bomb sniffing dogs are now commonplace in our government schools. How much are we willing to tolerate? Where will it end? More importantly, what is the root cause for these terrible changes?

Consider the tragic Columbine high school massacre that occurred in Colorado on April 20, 1999. One of the shooters, Eric Harris, admitted being motivated by the teachings of Charles Darwin. From his website, these insightful words were posted: *You know what I love? Natural selection! It's the best thing that ever happened on the earth, getting rid of all the stupid and weak.*

He and Dylan Klebold had planned that fateful event for over a year. They knew that day would be their last. Eric Harris wore a T-shirt displaying the most famous words of Charles Darwin. Emblazoned in large letters on the shirt was "**Natural Selection.**" Consider also the date of the massacre. It was not chosen at random. It coincided with the birthday of another infamous Darwin follower and Eric Harris' personal hero, Adolf Hitler. Ideas can have profound consequences for good or for evil.

Some of the survivors said before the victims were shot, they were asked, "Do you believe in God?" If they answered "yes," they were shot and killed. As with Adolf Hitler, this school shooter was trying to "help" the process of natural selection, by removing those he considered unfit, the stupid. He saw Christians as evil. It was especially shocking the media reports that saturated the airwaves for days following that fateful event failed to mention this obvious connection to Darwin and evolution. No doubt they knew, yet by their silence they misled the public they were supposed to serve.

School shootings are not limited to the United States. On November 7, 2007, an eighteen year-old student in Finland shot and killed seven students and the school principal before turning the gun on himself. The shooter, Pekka Eric Auvinen, died shortly after the shooting in a nearby hospital of self-inflicted wounds. As with the Columbine shooting, this event had been planned for months. Also like the Columbine shooters, Auvinen was a follower of Darwin. Graphic videos are posted on the Internet file-sharing website, *YouTube*. On a rambling text posted on his website, Auvinen said he was "a cynical existentialist, anti-human humanist, anti-social social-Darwinist, realistic idealist and god-like atheist. I am prepared to fight and die for my cause," he wrote. "I, as a natural selector, will eliminate all who I see as unfit, disgraces of human and failures of

257

natural selection." The song *Stray Bullet* accompanied Auvinen's posted video clip. It was from the rock band Kmfdm. The name of the band is a nonsensical and grammatically incorrect German phrase which loosely translated means, "No mercy for the masses." Both Eric Harris and Dylan Klebold, the Columbine school shooters, also cited that influential group's lyrics. Below are a portion of the lyrics from *Stray Bullet*. The entire song and others from the group can be found on the website: Kmfdm - Stray Bullet lyrics | LyricsMode.com.

Stray Bullet

I am your holy totem
I am your stick taboo
Radical and radiant
I'm your nightmare coming true
I have come to rock your world
I have come to shake your faith
Anathematic anarchist
I have come to take my place

I am your apocalypse
I am your belief unwrought
Monolithic juggernaut
I'm the illegitimate son of God
Stray bullet
From the barrel of love
Stray bullet
From the heavens above
Stray bullet
Ready or not
I'm the illegitimate son of God

From Kmfdm

On Sunday December 9, 2007, another copycat killer, 24-year old Matthew Murray, killed four people in Colorado. Two were killed at a mission training camp near Denver. Later the same day, two more people were killed at New Life Church in Colorado Springs before he was shot by an armed female church security guard. On an Internet posting, Murray said he intended to kill as many Christians as he could and admitted following the example of the Columbine shooters. He had two handguns, a rifle and over a thousand rounds of ammunition. He also confessed to being addicted to the same rock music advocating violence, as were the shooters at Columbine and Finland. Music, like ideas, can have profound effects for good or evil. Once again, an attempt was made to "assist" Darwin's natural selection with devastating results. As with Hitler and others, each of these murderers intended to hasten the natural selection process by removing those considered the unfit by the shooters. Ideas have consequences. Removing God, prayer, and the Ten Commandments from our public schools has had terrible consequences. Perhaps it is time to revisit those terrible decisions.

Life Without Purpose

The school shootings just described graphically illustrates the bitter fruits of applied Darwinism. Unfortunately, these are not isolated examples. Remember that evolution teaches human life is as nothing more than matter in motion. Man is but an animal. Only the fit should survive. It is believed the weak must be eliminated in order to improve the species. God and religion are mere myths of ignorant, superstitious man. Ethics and morality are not absolute, but are changing and relative.

Christian worldview has been largely replaced with nihilism. Foundational to the nihilistic view, there are no absolute standards or moral behavior codes. Everything is relative. Your rules do not apply to me. This is the rebirth of the '60's mantra, "If it feels good, do it." Sadly, public school students can no longer be taught ethics or morality for such teachings are not considered politically correct. Many parents have failed to pick up the mantle and teach their children right and wrong. Without direction, young people are adrift. They blindly follow popular culture seen in Hollywood movies or by rock stars. They lack purpose and direction. Evolution teaches life has no meaning. Despair often results. People, particularly young people, finding no purpose in life sometimes see no reason to live. When students are taught evolution as absolute scientific fact and that life is nothing more than matter in motion, a purposeful life quickly fades. Many become depressed, despondent. The nihilistic evolutionary worldview answers all important life questions in the negative. *Evolution is random and undirected . . . without either plan or purpose* (Biology, Prentice Hall, 1992). One of the consequences of this lack of purpose is the increase seen recently in teen suicides. Every three seconds, someone attempts to take his or her life and every forty seconds, someone is successful. The World Health Organization (WHO) now lists suicide as one of the three leading causes of death for people ages fifteen to forty-four.

Certainly the widespread acceptance of evolution is not the only cause of suicide, but such a philosophy may be a contributing factor along with the lack of parental involvement, low self-esteem and the availability of drugs. A senior lecturer and coordinator of complementary medicine at Monash University in Australia said: *We are more concerned with risk factors for depression, youth suicide, substance misuse and violence than the less-publicized protective factors, which include 'connectedness' and 'spirituality'.* – Craig Hassed, MBBS   Liberal syndicated columnist, Barbara Reynolds, said it best when she wrote: *One philosophy preaches happenstance with mayhem as a conclusion; the other, divine order. One suggests the survival of the fittest; the other, a commitment to serve the weakest and sickest among us.* (August 31, 1993)

All people, but especially young people, need to understand they are of great value in the eyes of God and that He has a plan and purpose for their life. *For God so loved the world that he gave his one and only Son, that whoever believes in him shall not perish but have eternal life.* (John 3:16) Such is the wonderful message of the Gospel. The message for Genesis is we were created special and in the image of the Creator. We are to be His image bearers. We are far more than mere "matter in motion" as the evolutionists proclaim. It is not difficult to see how the teaching that man is the result of purposeless evolution may exacerbate depression and low self-esteem among students and others with terrible consequences.

Darwin, Racism and Eugenics

The complete title of Darwin's revered book was *On the Origin of Species by Means of Natural Selection, or the Preservation of Favoured Races in the Struggle for Life.* It is obvious why the last phrase of the title is missing from most references of Darwin's book. That Charles Darwin was racist is known to most biologists, yet is seldom discussed. Certainly not all Darwinists are racists; however, Darwin's scenario of natural selection is premised on the notion of inherent inequality of the races. That is why evolution has been the basis for race-based eugenics programs and genocide throughout the world for more than a century. Evolution has been used to justify the killing of untold millions of people over the decades. Again, the general public is largely unaware race-based eugenics is still widely practiced today with terrible consequences. The reason such an unbiblical view is tolerated is obvious. Evolution proclaims man is not special or of more value to other animals. This perverse teaching is loudly proclaimed by today's secular culture. *We should come to terms with the idea that we belong to a highly specialized group of bony fishes* (Maisey, 1996). Nazism was a logical application of evolution by Hitler and other horrific dictators to "help" evolution along…to speed up the natural process by eliminating those deemed less fit. The results are well known. The consequences were dire with the murder of as many as 10 million Jews during the holocaust.

Charles Darwin, like Hitler and others after him, felt Europeans were more advanced than Africans and certain other races including the Aborigines of Australia. Racism certainly did not originate with Darwin for it has always been a part of fallen humankind. It seems to be an inherent human trait that somehow we see ourselves as better or more important when we put others down. My team is better than your team. What Darwin did was add scientific support and credibility to this oldest of human prejudices. Evolution fanned the fires of modern eugenics and genocide and has been carried to appalling extremes in many countries including the United States.

If you think racism is dead in the twenty-first century, consider a recent episode with one of the most respected and well-known scientists, Dr. James Watson, Nobel laureate and co-discover of the structure and importance of DNA in genetics. Shortly after his arrival in Great Britain on October 16, 2007, he made the statement at a news conference that western policies towards African countries were wrongly based on an assumption that black people were as intelligent as their white counterparts while testing results suggested the contrary. He felt the difference was genetic and the genes involved would be found and identified within the decade. Many were offended and shocked to find racism alive and well in the twenty-first century.

Modern eugenicists find support for killing the unborn in the earlier writings of Thomas Malthus. His groundbreaking publication, **An *Essay on the Principle of Population*** was published in 1798. His thesis was simple. If the human population continued to expand unchecked at the geometrically rate, humans would soon run out of food. He went so far as to predict this would occur by the middle of the nineteenth century. In spite of the failure of his dismal prediction to come to fruition, his work is still used to give scientific support to eugenics, abortion and genocide. Some consider Malthus to be one of the most important one hundred men to have ever lived. Charles Darwin was a life-long admirer of Malthus and referred to him as "that great philosopher" in a letter to J. D. Hooker in 1890. The struggle for human existence was the catalyst for Darwin's eventual application of the role of natural selection in the development of new species. Sadly, eugenics has been widely applied to humans throughout the world.

Joseph Needham, the founder of UNESCO (United Nations Educational, Scientific and Cultural Organization), with 193 active member countries, was strongly influenced by the writings of Malthus as was evolutionist Julian Huxley and his brother Aldous Huxley, author of ***Brave New World***. Karl Marx's socialist ideas were rooted in Malthus' theory as well. Malthus continues to have considerable influence. Perhaps the best-known example is Paul Ehrlich's ever-popular book, ***The Population Bomb***, published in 1968. It was required reading for many students and became an instant best seller, with his dire predictions about the uninhibited explosion of human population helped sparked the Green Revolution. It fanned the fires of eugenics and abortion during the last part of the twentieth century under the guise of scientific respectability.

Eugenics is often defined as, "The study of hereditary improvement of the human race by selective breeding." This definition sounds harmless, but the application of eugenic theory has caused incalculable death and human suffering. The United States Holocaust Museum has a portion of its website dedicated to those who were abused or killed because of Nazi eugenics, www.ushmn.org. The application of eugenics formed the primary motivation for Planned Parenthood. Margaret Sanger, the founder of Planned Parenthood was an outspoken advocate of eugenics as were those in association with her Birth Control League. Once again, we find racism at the core of eugenics. Dr. Hannah Stone wrote that, *the eugenicist, again, comes to birth control with a racial viewpoint. He sees in it an important aid towards controlling and improving the type and quality of the human stock.* We see proof of this in the current statistics for abortions with the rate of abortions for blacks far exceeding the rate for whites. Margaret Sanger confided in a 1939 letter to her friend Charles Gamble that, *We do not want the word to go out that we want to exterminate the Negro population.* To *purify* the breeding stock of the race at all costs remains the goal of eugenics. The American Birth Control League has as its slogan, *Fewer, but Fitter Children.* How the media can continue to keep such racist aspirations from the public is incomprehensible.

Bill Gates, chairman of Microsoft, is one of the world's wealthiest men and widely respected for his groundbreaking work in computer programming. He is considered the leader in the personal computer revolution. He is also a strong supporter of Planned Parenthood. His father was at one time the head of Planned Parenthood. Bill Gates admits to being strongly influenced by the writings of Malthus. Like so many others today, he sees abortion as an effective way to reduce the growth of human population.

In many ways, the early eugenics in the United States paralleled that of Nazi Germany. The U.S. doctors believed sterilization could forever rid society of mental illness and crime. A recent study by Yale University found these once-popular movements aimed at improving society through state-authorized sterilizations were conducted longer and on a much larger scale in the United States than previously thought. Beginning in 1907, over sixty thousand Americans were sterilized, many without their knowledge or consent. The eugenics program was a logical outgrowth of the philosophy of social Darwinism, which envisioned the human society in terms of natural selection and suggested science could engineer and speed the process by attacking such assumed hereditary problems as moral decadence, crime, venereal disease, tuberculosis and alcoholism. Both German and American eugenics advocates believed science could solve the existing social problems with compulsive sterilization of the less fit. The practice in the United States did not end until the 1960's, after repeated court challenges and the growing influence of the civil rights movement. Few today are aware of how recently forced sterilization was practiced in the "land of the free."

While the Nazi claims of Aryan superiority are well known, researchers in the United States thought sterilization could help avoid being overrun by the influx of "lower races" from southern and Eastern Europe. The U.S. eugenics program drew glowing reviews from authorities in pre-Nazi Germany. Both programs were respected by leading scientists of the day. From the editors of the prestigious New England Journal of Medicine in 1934 we find, *Germany is perhaps the most progressive nation in restricting fecundity among the unfit.* Supreme Court Justice Oliver Wendell Holmes wrote in the majority opinion of a landmark eugenics case in 1926, *It is better for all the world, if instead of waiting to execute degenerate offspring for crime, or to let them starve for their imbecility, society can prevent those who are manifestly unfit from continuing their kind.* Compulsory sterilization programs were seen as attempts of genetic engineering and were focused disproportionately on the poor and disenfranchised groups. Nazi Germany sterilized over 400,000 in the 1930's. Sweden sterilized 62,000 from the 1930's to the 1970's. In the United States, release from mental institutions were often predicated on forced sterilization and Native Americans were often sterilized against their will while in the hospital for other reasons such as childbirth. Thirty-three states had compulsory sterilization programs. History clearly points to Darwin as providing a scientific veneer to prevailing racism. The biblical record of creation provides the strongest proof of the true equality and value of humankind and of hope for all eternity. *So it is written: The first man Adam became a living being; the last Adam, a life-giving spirit. The spiritual did not come first, but the natural, and after that the spiritual. The first man was of the dust of the earth, the second man from heaven.* (1 Cor 15:45-47, NIV)

Abortion

Abortion has its roots deep in Darwinian soil and is a logical outgrowth of the Darwin inspired eugenics program in the United States in the first half of the 1900's. Planned Parenthood's founder, Margaret Sanger, advocated the removal of human *waste* and was in favor of racial genocide as the elimination of human *weeds*. Today, some of Planned Parenthood's leaders suggest it is a good thing to eliminate *useless eaters*. Consider carefully a philosophy that some people are thought of as mere *human weeds* and *useless eaters*. Contrast this with God's unwavering sacrificial love of *all* people.

Abortion is the most tragic and far-reaching application of Darwin's concept of evolution. Worldwide approximately 46 million abortions are performed each year! The abortion rate in the United States is higher than that in Australia, Canada the UK and other western European countries. Nearly half of all pregnancies among American women are unintended, in spite of the availability of birth control and sex education. There remains a huge racial difference. Pregnancies are terminated by abortion in only 40 percent of white women compared to 69 percent of black women. The total number of abortions in the United States since 1973 is over 50 million. This number approaches those murdered by both Hitler and Stalin. The connection to evolution is obvious. People are taught man is but an animal and has no intrinsic value. The unfit, even the unwanted should be eliminated, without consequent or guilt. In spite of what we often hear about the need for a woman's right, the mother's health, rape and incest result in fewer than 7 percent of all abortions.

As a former university professor, I taught nursing students for many years and each class had women that had aborted one or more unborn children. I conducted a very open class and we discussed such things. It broke my heart to hear many abortions were carried out simply because "the timing was wrong" or they "wanted a boy, but I was carrying a girl." Perhaps the most revealing excuse heard more than once was, "My husband had a vasectomy and would divorce me if he found out I was pregnant." An innocent child killed for the sins of the mother. One is reminded of the clear teaching from scripture. *The LORD is slow to anger, abounding in love and forgiving sin and rebellion. Yet he does not leave the guilty unpunished; he punishes the children for the sin of the fathers to the third and fourth generation.* (Num 14:18, NIV)

Apart from the senseless killing of the unborn, abortion is big business in the United States…a business supported financially by our government. *It's an outrage that our government has subsidized an abortion performing organization such as Planned Parenthood to the tune of more than $3 billion over the past four decades,* said Jim Sedlak, executive director of American Life League's STOPP International. *Planned Parenthood's 2002-2003 Annual Report shows that 33 percent of its income came from government grants and contracts totaling $254.4 million in the fiscal year ending in June 2003. That's a 5.6 percent increase from 2002. This government money helped Planned Parenthood haul in a $36.6 million profit in its last fiscal year.* Our tax dollars have been spent without our permission to kill the innocent. Sadly, many in congress and President Obama want to increase government funding of abortion worldwide and overturn many of the court decisions and state laws protecting the unborn in the United States.

Darwin inspired Genocide

Historians generally agree the three most sinister people of the twentieth century are Adolph Hitler, Karl Marx and Joseph Stalin. Directly or indirectly, they are attributed with the murders from 50 to as many over 100 million innocent people. They also played a major role in causing the worst war the world has known. Causality estimates from the Second World War range from 50 to 70 million people. What is missing from the history books is these men were directly or indirectly influenced by the teaching of Charles Darwin's evolution by natural selection. Let's consider Darwin's influence on these dark figures in some detail. We will see Darwin's evolution as a common denominator for mass murders past and present. This is the deadly legacy of Darwin.

Karl Marx (1818-1883) wrote the **Communist Manifesto** in 1848 and is considered the father of the communist system. He is widely known as an avid adherent of Charles Darwin's theory and applied evolutionary principles to social and economic ideas. Darwin claimed animal species were not doomed to remain as they were, but could evolve into something better. Karl Marx applied this theory to society and introduced the idea of "scientific socialism" claiming social classes were in conflict with the higher classes exploiting the lower classes. He predicted the exploited classes would rise up and overthrow the dominant class and the result would be a classless society with everyone equal and no need for government. Atheism is a basic tenet of Marxism in general and Soviet Communism in particular.

Evolution was an important corner stone in the birth and growth of communism. Lenin, Trotsky and Stalin were all atheistic evolutionists. Marx wrote Engels and said regarding The **Origin of Species** that this book contains the basis in natural history for our views. They saw it as the biological counterpart to class war. Darwin provided the key agent for bringing about the ideological changes for fostering the growth of communist theory. Marx's philosophy, like that of Hitler reflected the brutality of nature graphically described by Darwin. Clearly, some of the most inhuman atrocities ever committed have been perpetuated in the name of national socialist or Marxist "social advancement" based on evolutionary philosophies. (Ankerberg and Weldon, 1998)

Vladimir Lenin (1870-1924) was the founder of the Russian Communist Party and first head of the Soviet State. He was a disciple of Karl Marx and applied the evolutionary view to justify decades of utter ruthlessness and terror in Russia. The term, "rivers of blood" is used in describing his reign of terror. He masterminded the bloody Bolshevik rise to power in Russia that began in October of 1917. Lenin convinced the Bolshevik Party to form an insurrection against the Provisional Government and announced his attempt to construct a utopian socialist order in all of Russia. The former Soviet Union was born. Lenin's preserved body is on permanent display at the Lenin Mausoleum. Over time, his character was elevated to the point of near religious reverence. By the 1980's, every city in the Soviet Union had a statue of Lenin in its central square. Collective farms, medals, wheat hybrids and even an asteroid were named after him. Nursery children were taught stories about "Grandpa Lenin." (Source: http://en.wikipedia.org/wiki/Lenin, December 14, 2007.)

Leon Trotsky (1879-1940) was second only to Vladimir Lenin in the early Soviet communist rule. He was fanatically committed to Darwinism and

Marxism and became the brutal enemy of the Christian church throughout Russia. Trotsky said Darwin's ideas "intoxicated" him and that Darwin *stood for me like a mighty doorkeeper at the entrance to the temple of the universe.* Without the laws of the Creator-God, he felt unstrained to use any means to attain power and political ends, again with devastating consequences. He was founder and the first commander of the Red Army. He was also among the first members of the Politburo or Presidium, the Central Committee of the Communist Party of the Soviet Union. (Source: http://en.wikipedia.org/wiki/Leon_Trotsky, December 14, 2007)

Adolf Hitler (1889-1945) was leader of the National Socialist (Nazi) German Worker's Party. He was appointed Chancellor of Germany in 1933 and became the "Fuhrer" in 1934, remaining in power until his suicide in1945. He launched World War II and for that alone bears the responsibility for the death of millions. He clearly formed his racial and social policies on the evolutionary ideas of survival of the fittest and the superiority of certain 'favored races' (as in the title of Darwin's book). Hitler's reign resulted in the murder of six million Jews as well as many blacks, gypsies, those deemed mentally retarded, and other groups he considered unfit to live. The evolutionary 'science' of eugenics provided him with justification for his terrible racial genocide.

Joseph Stalin (1879-1953) is generally agreed to have been the most powerful and murderous dictator in history. Historians disagree regarding the influence of Darwin on Stalin. There was a Soviet claim that he read ***The Origin of Species*** at the age of thirteen and told a pupil it proved God does not exist. There is doubt today this actually occurred in part because Stalin remained religious, even pious for several years after this alleged reading of Darwin. Nevertheless, in later years he was an ardent evolutionist and the supreme ruler of the former Soviet Union for a quarter of a century. He understood that evolution provided no basis for conscience or morality. He felt free to torture and murder to whatever extent was needed for him achieve his communist goals and he did. He surpassed even Hitler in his zeal to remove the unfit and is thought to have murdered at least ten times as many "inferiors." Estimates range from 60 to 100 million people murdered to "improve the species."

Pol Pot (1935-1998). The death of Cambodia's marked the end of one of the world's worst mass murderers. From 1975, he led the Khmer Rouge to genocide against his own people in a bloodthirsty regime, which was inspired by the communism of Stalin and China's notorious Mao Zedong. Chairman Mao is known to have regarded Darwin and his disciple Huxley as his two favorite authors. The combined effects of slave labor, malnutrition, poor medical care and rampant executions (http://en.wikipedia.org/wiki/Pol_Pot) caused the estimated death of nearly 2 million people or 25 percent of the entire population. (Source: , December 15, 2007)

Worldviews At War

Everyone has a philosophical outlook on life, an articulate defense, or more commonly, a vague notion regarding his or her origin and destiny. Such beliefs are called our worldview. Today we find an increasing polarization between those holding a biblical worldview and those that do not. Such conflicting views fan the fires of discord regarding such diverse topics as creation science and evolution, ID, animal rights, abortion, morality, national and local politics and countless other societal issues. Everyone agrees the war of the worldviews is heating up with no end in sight. As a side issue, it is no longer considered politically correct to discuss racial matters. That changes nothing. Some people cling to racial prejudice and our inability to openly discuss them only makes matters worse. Discussion can lead to a better understanding and mutual respect. It is the same in science.

I have known Dr. Jerry Bergman and his writings for many years. I had the pleasure of meeting him at a board of directors meeting for the Creation Research Society nearly thirty years ago shortly after his own tenure denial at Bowling Green State University. We have stayed in touch, sometimes reviewing each other's writings. Dr. Bergman wrote an insightful review of some of the events surrounding his tenure denial and dismissal. By all accounts, he was fired for writing a forty-five page monograph entitled, *Teaching about the Creation/Evolution Controversy.* It was widely distributed and is still available in over six hundred libraries worldwide. The book was totally objective; he neither defended nor attacked evolution or creation. He simply wrote about how to teach about the controversy. He received hundreds of letters about the book and many were very positive and appreciative. Still, some in academia attacked him viciously for even mentioning the forbidden subject. His case went to court in 1985 and his tenure denial was upheld as happens about 88 percent of the time. He was devastated and had to literally start over eventually earning nine degrees. He is a prolific author publishing over seven hundred publications in twelve languages plus twenty books and monographs, many dealing with the creation/evolution controversy. For more details about Dr. Bergman, read his own story in chapter 14 about religious persecution. Sadly, his case has been repeated many times over. The battle continues unabated today.

Double Standard of Society

One does not need to surf the Internet long to encounter a plethora of anti-creation and anti-Christian websites. This is strange in light of society's insistence of the importance of tolerance, political correctness with its alleged need for sensitivity toward the viewpoints others. One website attempts to expose the Christian faith as totally unreasonable, aggressive, and intolerant. Other websites contend scripture as little more than errors, contradictions, and impossibilities. Obviously such cynical musings do not align with the real world. Recently, Brad Stetson reviewed a book by A. J. Schmidt, **Under the Influence: How Christianity Transformed Civilization.** Stetson summarizes Schmidt's book by saying it is a comprehensive and remarkably thorough investigation into all that Christianity has given human civilization. Foremost among these gifts is respect for people. From saving throwaway babies of the Roman Empire, to opposing abortion, to caring for the sick, elderly and handicapped, it was Christianity that protected the weak and made it possible for human equality to become a reality. Other contributions include the invention of hospitals in the fourth century, as well as orphanages and charitable organizations. Further, the Christian inception of universities gave us the basic concept of higher learning that has contributed so much to human progress. This is certainly true when it comes to the nature of science. In **The Soul of Science,** by Pearcey & Thaxton (1994), the authors quote the writing of evolutionist Loren Eiseley who stated that science *demands some kind of unique soil in which to flourish.* What is that unique soil? Eiseley identifies it, somewhat reluctantly, as the Christian faith. *In one of those strange permutations of which history yields occasional rare examples,'* he says, *it is the Christian world which finally gave birth in a clear, articulate fashion to the experimental method of science itself.'* Eiseley is not alone in observing that the Christian faith in many ways inspired the birth of modern science. Science historians have developed a renewed respect for the middle ages, including the Christian worldview, culturally and intellectually dominant during that period. Today, a wide range of scholars recognizes Christianity provided both intellectual presuppositions and moral sanction for the development of modern science.

One may wonder why atheists spend so much time and effort attacking the Biblical faith of millions while all but ignoring other beliefs. This comment was also found on one of the many anti-creation websites on the Internet: *First of all, I am not against Christianity. In fact, I am a Christian . . . What I am opposing here is the secularization of our public schools under the guise of so-called Creation science . . . it is pseudo-science; it is fundamentalist, radical, dogmatic, theocratic and self-justifying religion. In particular, I am opposing the specific doctrine of Young Earth Creation science.* This Christian brother need not resort to name-calling. Instead, let him list some empirical evidences for neo-Darwinian macroevolution to make his case. Such a summary would silence the "Young Earth Creationists." Instead, let us apply his list of names to the alternate view, that of evolution, as viewed by its leading thinkers.

Pseudo-science?

"We take the side of science *in spite* of the patent absurdity of some of its constructs, *in spite* of its failure to fulfill many of its extravagant promises of health and life, *in spite* of the tolerance of the scientific community for unsubstantiated just-so stories, because we have a prior commitment, a commitment to materialism. It is not that the methods and institutions of science somehow compel us to accept a material explanation of the phenomenal world but, on the contrary, that we are forced by our *a priori* adherence to material causes to create an apparatus of investigation and a set of concepts that produce material explanations, no matter how counterintuitive, no matter how mystifying to the uninitiated. Moreover, that materialism is an absolute, for we cannot allow a Divine Foot in the door" (R. Lewontin, "Billions and billions of demons," *The New York Review*, January 1997, p.31).

Fundamentalist?

"There was a world of bacteria here that preceded us" (Kenneth Nealson, microbiologist at NASA. *Newsweek*, September 21, 1998, p. 12). "Evolution is a process which has produced life from non-life, which has brought forth man from an animal, and which may conceivably continue doing remarkable things in the future. In giving rise to man, the evolutionary process has, apparently for the first and only time in the history of the Cosmos, become conscious of itself" (T. Dobzhansky, "Changing Man," *Science*, v. 155, January 27, 1967, pp. 409–15).

Dogmatic?

"Evolution is a fact, fact, fact!" (Michael Ruse, *Darwinism Defended*, 1982, p. 58).

Radical?

The Kansas Science Education Standards (December 1998) states that these "unifying concepts and processes transcend the traditional disciplines of science." An anthropologist at Rutgers contends that "Darwinian science inevitably will, and should, have legal, political, and moral consequences." (*Scientific American*, October 1995)

Theocratic?

"Like all important ideas, evolution attracts controversy. It has affected not only science, but also philosophy, religion, and human attitudes" (*Botany*, W. C. Brown, 1995). As is easily seen, naturalistic evolution is a total worldview, not well based in real science. Name-calling doesn't change this fact.

Lack of Objectivity in Teaching Science.

The Spring 2003 issue of the *National Science Teachers Association* recommends an anti-creation book, authored by three evolutionists entitled, *The Creation Controversy & the Science Classroom*. Talk about saber rattling! In the single paragraph that extols this surprisingly brief (64-page) book, confrontational words such *as opposition, debate, ammunition, forceful, arms and strategies* are found. Ironically, a quote from an elementary school teacher in Cabot, Pennsylvania, on the same page says the book was written in "neutral terms"! It would seem that the secular community's right hand doesn't know what the left is doing. On one hand, books such as the above are written to formally *condemn* creation science in public school classrooms, while at the same time evolutionists proclaim it's "unconstitutional" to *teach* creation science in public school classrooms! Thankfully, one of America's foremost censors of creation science admitted, "The Supreme Court decision says only that the Louisiana law violates the constitutional separation of church and state; it does not say that no-one [sic] can teach scientific creationism — and unfortunately many individual teachers do." (Scott, E., Correspondence in *Nature*, 329 (Sept. 24, 1987): 282)

The origins issue will continue to be a battlefield because evolutionism is not just a theory of biological origins, but the basic foundation of the secular worldview. There are no living sciences, human attitudes, or institutional powers that remain unaffected by the ideas released by Darwin's work.[2] (Collins, J., a philosopher quoted by Miller & Levine, *Biology*, 1995, p. 313). Secularists understand how important this battle of the worldviews is, much more so than do most church members. The late S. J. Gould said, "Modern creationism, alas, has provoked a real battle" (Gould, S.J., *Rocks of Ages*, New York: Ballantine Publishing Group, 1999, p. 125) and "This battle must be won." (Alters & Alters, *Defending Evolution*, Jones & Bartlett Publishers, 2001, p. 4). But battle objectives are confused by atheists such as Niles Eldredge who recently said "[Creationists] are motivated primarily to see that evolution is not taught in the public schools of the United States." (Eldredge, N., *The Triumph of Evolution*, W.H. Freeman & Co., 2000, p. 11).

This is an erroneous premise, of course. Creationists do not advocate removing evolutionary teaching in public schools. We would, however, like the many *scientific* problems regarding evolutionism clearly addressed in the free marketplace of ideas. We would attempt to present to young people in our tax-supported public schools a non-Biblical origins model alongside the questionable science of evolutionism. Advocates of critical thinking skills could only agree to such a suggestion, and students on both sides of the issue would benefit. The fact is that creationism is no longer [only an] American problem," says Michael Zimmerman, professor at Butler University in Indianapolis, architect of The Clergy Letter Project, an alliance of Christians who back evolution. "The best way to overcome this pernicious situation is for religious leaders and scientists [of course, by this he means only evolution-scientists] to come together to discuss how religion and science [only evolution-science] can be compatible - how they use different methodologies to help people understand the world and the human condition."   Watch out for Zimmerman, et al. His big boss at NCSE www.ncseweb.org , Eugenie Scott said, "I have found that the most effective allies for evolution are people of the faith community. One clergyman with a backward collar is worth two biologists at a school board meeting any day!" (*A Conversation with Eugenie Scott*, Science & Theology News, 4/1/02 ; www.stnews.org/Commentary-1835.htm

Rampant Pantheism

Pantheism is described as any system of belief that includes the teaching "God is all" which is characterized by many Hindu and Buddhist doctrines. Pantheism, in other words, is identical with the whole natural world composed of forces, substance, and laws evident in the universe.

Many of those holding to animal rights, which more often than not is based on pantheism, are increasingly vocal and insist that the lines between you and your dog be eliminated. The philosophy of evolutionism supports this strange idea. As man continues to be denigrated as "a virus infecting Earth" by some environmental groups, his pets and non-domestic creatures are enjoying a new-found esteem. This is thanks, in part, to bizarre activist groups such as In Defense of Animals (IDA) and "animal-rights" law courses being taught at institutions like the prestigious Rutgers School of Law. Indeed, the San Francisco IDA would like to see the word "guardian" used in reference to pets rather than "oppressive terms such as 'owner' or 'master.'" One litigator and animal rights law professor went a step further calling for a granting of personhood to baboons and chimpanzees! Philosopher, Pete Singer, maintains orangutans, gorillas, and chimps should have legal equality with man.

Another interesting story involves a legal organization representing prisoners. This "civil libertarian" group is currently engaged in a fight regarding the alleged harmful effects of chaining inmates together during work hours because it makes the convicts feel like animals. Why the fuss? Have not these malefactors and other members of society been constantly taught in public school that they are animals? We see in taxpayer-funded high school biology textbooks that people are but a branch of the phylogenetic tree, the result of many millions of years of 'descent with modification.' "Human DNA represents millions of years of accumulated evolution," asserts evolutionist James Trefil (*101 Things you don't know about science* 1996, p. 270). In 1998, Bruce Alberts, President of the National Academy of Sciences stated: "In the last ten years we've come to realize humans are more like worms than we ever imagined." How can being chained together with other people be any more demeaning than being told outright that we have worms as ancestors? On one hand, we see serious legal efforts to have people treat their animals virtually as equals and grant chimps personhood, while on the other hand, special interest groups are outraged because men are treated like animals. This quandary sounds similar to the child's dilemma: you can't have your cake and eat it too.

Compromise Within the Church

As we have all seen, pastors take advantage of special occasions to preach biblical truths and present sermons of importance to remember God's blessings upon His people. For example, besides Christmas, Easter and our Nation's birth, we have "Sanctity of Life Sunday," intended to remind Christians of the millions of unborn children who have been murdered since 1973 through abortionists and their helpers. Sadly, not everyone is aware that every February 11 has been declared "Evolution Sunday" by over ten thousand ministers across the United States. The community of theistic evolutionists ("God created evolution") has undertaken this day to confront those who accept the Genesis account of Creation. They propose to worship Saint Darwin in the sanctuary on a given Sunday every February. Initiated by the blasphemous "Clergy Letter Project," this day of remembrance coincides with the birthday of Charles Darwin, the father of modern evolutionary thought. The web site of the *Clergy Letter Project* has announced:

On each February 11, hundreds of congregations from all portions of the country and a host of denominations come together to discuss the compatibility of religion and science. For far too long, strident voices in the name of Christianity have been claiming that people must choose between religion and modern science. More than ten thousand Christian clergy have already signed The Clergy Letter demonstrating that this is a false dichotomy.

Evolution Sunday could just as well be called "Compromise" Sunday. On their website, they present the tired and false picture of "science vs. religion" that institutions such as ICR and CRS allegedly tout. To set the record straight, neither CRS nor ICR sees science as the enemy. In fact, all faculty members at ICR Graduate School have their advanced degrees in scientific disciplines from accredited, secular institutions (see ICRGS Faculty). The same is true for voting members of CRS. The accusation that we somehow fear science or find it opposed to Scripture is patently false. What we do teach is that the philosophy of macroevolution (i.e., the unobserved past processes of vertical evolution) is at odds with Genesis.

Scientists who believe Genesis are not the only experts who balk at the idea of evolution. Creation scientists believe that all science, correctly understood, is perfectly compatible with Scripture. The great scientists of the past such as Newton, Kepler, Boyle, Maxwell, Pasteur, Mendel, Faraday, and many others, believed the same. If the conduct of genuine science is not the issue, then what is? The notion of a Creator is at the root of these attempts to bring God's Word down to the level of man's theories.

Let's remember that Darwin's theory of evolution was dreamed up as a secular explanation for the origin of people, plants and animals. It was never intended to be interpreted within a biblical framework. Darwinism was, and is, popular because it attempts to explain "creation" without a Creator. These ten thousand clergy members will sadly be using their pulpits to pound a square peg into a round hole. Good science is at odds with the strange notion of macroevolution, but in total agreement with Genesis.

In summary, the fruits of evolution have been widespread and devastating. Murderous dictators, past and present, have justified their atrocious acts firmly on godless evolution. Even today the unborn are killed in numbers exceeding those killed in all the wars of history because evolution teaches humans have no value and there is no life after death or judgment. Sadly, some pastors have joined the mantra. As concerned Christians, we must first be informed ourselves and then tell others of the extreme devastation evolution has wrought globally and here at home.

Chapter 13
Implications for Science

But ask the animals, and they will teach you, or the birds of the air, and they will tell you; or speak to the earth, and it will teach you, or let the fish of the sea inform you. Which of all these does not know that the hand of the LORD has done this? (Job 12:7-9, NIV)

Science is an important part of our God-given responsibility to have dominion over the earth. We are to study nature and in doing so, see the hand of the Lord as many of the great scientists in the past have done. Yet, many today see the world through the tainted lens of materialistic evolution. Modern secular science tells us we live in a totally natural world brought about by random events and mutational mistakes over vast periods of time. If, instead, an all-knowing and loving Creator designed the world and created life with all its beauty and complexity, then everything changes. The entire paradigm shifts. Scientific research based on an erroneous evolution model will be unfruitful and wrong conclusions can and have been drawn. Time and research funds will be wasted. Even more importantly, due to the pervasive materialistic mindset scientists no longer have the freedom to follow the evidence wherever it leads. If this continues, science as we know it will wither and die. Something must change.

There is hope. In many ways it is *not* a good time to be an evolutionist. After 150 years of diligent searching for scientific support of Darwin's theory, no evidence has been found supporting macroevolution. Evolution is approaching intellectual bankruptcy and thousands of scientists have already abandoned macroevolution like deer fleeing an approaching forest fire. The origin of life and the complexity of living things remain an unsolved mystery and objective scientists are finally beginning looking elsewhere. Yet the dogma persists. Many still blindly cling to evolution by faith. For them, it has become a religion. True science began in Genesis when God gave mankind dominion over the newly created world. Many leading and productive scientists of the past followed this model with huge success. Science must return to this foundation. Ignoring this basic tenet will continue to lead many scientists down a fruitless path.

Life did not evolve; it was created. Man is not the product of mindless genetic errors, but was supernaturally created with purpose and inestimable value. Heaven, hell and judgment are real. Life after death is an unavoidable reality. There are eternal consequences for our decisions and our behavior. The beauty and complexity clearly evident in living things unequivocally points to design and the Creator instead of blind chance. Indeed, *"the heavens declare the glory of God."*

Let's assume for a moment the Watchmaker is *not* blind and the Bible is trustworthy. Let's assume for a moment the biblical account of origins as recorded in Genesis and throughout scripture is valid. What are the scientific implications? There are many. Following are a few examples. Many others could be given and still more will come to light in the near future as human knowledge continues to increase. Science will be more productive if scientists can again be free to use the Creation worldview as was done throughout most of the history of science. The materialistic approach to science has led many astray and in the area of the origin of life is totally bankrupt. Thoughtfully consider the following examples. A change of direction is sorely needed for science to survive.

No Vestigial Organs

If man was created by an all knowing Creator, then our bodies do not contain useless organs. All of our parts have function and are useful. This seems simple, but has far reaching implications. As shown previously, there was at one time a list of nearly two hundred organs thought to be useless relics of our animal ancestors. As we learned more about human physiology, the shorter the list became until today there are none. If creation is true, we must study to better understand the human body and how each of the parts function. When the function of any organ is questioned we must assume it was created for a purpose and search for its purpose. Scripture demands this approach and the facts of science now support this view as well. There is a recent example from molecular biology. With the elucidation of the human genome, it was estimated as much as 97 percent of our DNA had no function and was termed "Junk DNA." Once again, as knowledge is replacing ignorance, it seems even the so called "Junk DNA" is important and has many varied functions in man and beast.

Of course the same is true for all living things. When the function of a structure or behavior appears unknown, let this be a red flag indicating additional research is needed. We must no longer look blindly into the alleged past evolutionary history and make false allegations of the structure's past value in some assumed ancestor. Only with this mindset can true functions be discovered and science advance. We have been chasing vestigial ghosts for far too long and the results have been spooky and not scientifically productive. Countless thousands underwent the risks and discomfort of needless appendectomies and tonsillectomies.

Phylogenetic Trees

Countless pages of biology textbooks and endless hours of lecture are devoted to attempting to show how all plant and animal species are related to a common ancestor. Speculation abounds for there is no actual supporting evidence. This concept is central to all of biology, yet such endeavors are based on faith in evolution dogma, not facts. Even our current taxonomy or the classification system for living things assumes kinship and common descent. If, instead, all things were created and only limited changes have occurred, this time can be better spent in more productive areas. Assumed phylogenies are but fairy tales for the unlearned and are used to prop up a failed theory. Even as a student, I was amazed at the lack of actual evidence supporting the endless phylogenetic trees and resented the wasted time. The gaps in the fossil record are real and will continue to exist as Darwin himself feared. Indeed, something is strangely amiss. A new approach is sorely needed. For far too long, there has been unsupported empty chatter and speculation.

If all the kinds of living creatures described in Genesis were created much as they remain today, then the original "kinds" need to be defined and the extent they can and have changed over time needs to be determined. The only truly useful phylogenetic trees would be those based on the originally created kinds described in Genesis. Limited variation is what is seen today, not the magical transformation from a lizard to a bird as demanded from macroevolution. This opens an entire new and more productive area for study in phylogeny. Such studies are already underway by creation scientists and this work should be applauded, funded and expanded. A new classification system is needed and will better fit actual living things. Some Christian taxonomists have started this ground breaking research, but much more work is needed. Science must return to its scriptural foundation to advance. As Sherlock Homes famously said, "Just the facts Watson, just the facts."

Embryonic Development

Perhaps no area of science has so much been built on so little evidence as in the area of embryonic development. We have all seen the drawings of embryos in various stages of development with the earlier stages looking more and more alike. These drawings have been reproduced countless times in various high school and university textbooks. Yet, we have known for nearly 150 years that the original drawings were fraudulent. Early embryos of various animals are actually quite distinct. This lie seems to have taken on a life of its own and all of embryology needs to be reconsidered and the differences openly discussed and studied. Certainly all complex animals begin life as a single fertilized egg, but beyond that, many differences are evident and those differences must be studied and the reasons for the differences be understood. Even if evolution were true, there is no rational reason for embryos to replay their assumed evolutionary history.   As mentioned before, the human lungs develop late in embryonic development and in their final form. This would suggest we evolved recently from animals lacking lungs. Again, I actually saw the humor in this as an undergraduate student and brought it up in class. The students appreciate it far more than did the professor.

A new approach in embryology is desperately needed. This is especially true in light of advances in molecular biology and genetics. The language of DNA must be interpreted as being written by the Creator and not the result of endless errors over time. We now understand God as Author of every living thing and the language He used was DNA. We are at the dawn of vastly increasing our knowledge of genetics in and finding cures or treatments for various genetic disorders.   Let us see DNA for what it truly is: the handwriting of God our Creator.

Fossils and the Flood

Once again let's assume the biblical account of the global flood actually happened as described in Genesis. This alone accounts for the vast fossil record evident in sedimentary rocks around the world. All of paleontology needs to be re-examined in light of scripture. Once again, this changes everything. *The Genesis Flood* was an excellent beginning. It is time now to continue that work. Let's "speak to the earth" and listen closely at what it has to say. Again, there are pockets of Creation scientists doing this very thing today. Let's encourage and fund such activities. There needs to be far greater organization, communication and cohesiveness. International conferences must be conducted and information shared. Again, paleontology will be more productive and of far greater value when we begin looking for evidences of the originally created kinds instead of wrongly assumed common ancestry.   There could be no stronger proof of the global flood described in Genesis than the vast fossil record. Paleontologists must again interpret their findings in that light as was done in times past. That time is now and the stakes are high.

New Light in Astronomy

Astronomy, like biology, needs to be seen as the result of a Creator-God and not a mindless Big Bang. One recent powerful book and a revealing movie address these important issues and have provided a truly outstanding leap from past astronomical fiction. *The Privileged Planet* provides convincing evidence the earth was not only created special to support life including man, but its placement in the universe was ideal for scientific discovery. From the complex world of atoms and molecules found within living cells, to the vast reaches of space the hand of the Creator is evident, if we but look.

The other provocative movie, *The Star of Bethlehem*, shows the one time alignment of the stars and planets to announce the birth of Christ as revealed in scripture. Again, the wisdom of Creator in announcing the birth of His Son is finally understood and speaks volumes of His wisdom and the supernatural alignment of the heavenly bodies at the very time our Savior was born, and again while he died the terrible death on the cross the stars had much to say. Indeed once again we find, "the heavens declare the glory of God." Let's begin to look at this truth in new light. Much more is to be learned in this and related areas.

Celebrating the God of Creation

The most exciting prospect of this new approach to science is not only more scientific discoveries, but nature will once again be seen as unmistakable evidence of God's power and wisdom. It is for this reason we have hundreds of Bible passages praising the God of Creation. I can hardly contain my excitement when I contemplate some of the many ways science too can praise the Author of all things. I pray I will live to see the day when PBS and other nature programs look at nature and see the hand of God in the beauty and complexity so evident. Let us praise Him alone for, *Through him all things were made; without him nothing was made that has been made.* (John 1:3, NIV) Amen and amen!

Swamp Rabbit.

Chapter 14
Christian Persecution in America Today

In fact, everyone who wants to live a godly life in Christ Jesus will be persecuted (2 Tim 3:12). If the world hates you, keep in mind that it hated me first. (John 15:18)

When someone mentions Christian persecution or martyrdom, most of us think of the terrible trials experienced by the first century Christians. Indeed, thousands of early Christians died painful public deaths for their convictions. What most people do not realize is that untold thousands of Christians are persecuted today and many killed for their faith. According to Voice of the Martyrs (www.persecution.com) and other reliable sources, approximately 500 Christians throughout the world are killed every day for espousing a Christian worldview. We do not learn this in school, nor is it reported on the evening news. Think of it, more than 180,000 Christians killed each year and we hear nothing! It was widely reported when we lost 1,000 service men and women in our war on terrorism, yet nothing is said as thousands are killed every month for taking a stand for Christ. Shockingly we need not travel to a far away land to see Christians persecuted today. It is happening right here in America.

Most of us learned in public schools that the United States was founded in part for religious freedom. We understood our constitution assures freedom of speech and religious freedom for all. At least that was the case until history revisionists began rewriting our history books. It is sad and shocking that right here in the United States, where we boast religious freedom, we have professionals losing their jobs and careers due to religious persecution. The coliseums of old have been replaced by the ivory towers of academia, the lions by university administrators.

Over three thousand highly educated and experienced university professors have been denied tenure are outright fired for rejecting evolution. That number continues to grow. For most of them, this tragic event marks the end of their professional careers. Many must seek employment outside their area of interest and training because the academic doors have been rudely slammed in their faces. Most are black listed by the ACLU and other anti-Christian organizations making future academic employment impossible.

Perhaps even more tragic, hundreds of intelligent and highly motivated young graduate students have been expelled from educational programs in the life sciences for doubting evolution. They have been barred from higher education, while being taught science is objective. Again, it seems the media is unwilling to report this ongoing travesty of justice. The public is largely unaware of this modern form of religious persecution, yet it abounds and is becoming more and more commonplace each year. Let's consider a few case studies. Obviously this topic is near and dear to my heart.

E. Norbert Smith, Ph.D.

Let me begin this chapter with the account of my own tenure denial. As with so many others, it marked the end of my professional teaching/research career in academia and I returned to the family farm of my childhood. Unable to support my family farming, I worked as an oil field roughneck then ended up teaching electronics in a federal prison. Eventually I worked at a small community college teaching microbiology and electronics courses. Unable to make enough money for retirement from teaching, I became a truck driver and drove 18-wheelers for twelve years. Let me share my story.

The job market is always tight for graduating zoology students. A little known academic fact is that many students who continue their education by accepting a postdoctoral position do so because more lucrative permanent teaching/research positions were unavailable. They applied for teaching positions, but failed to get an offer. Like most last semester life science doctorial students, I applied for over two hundred teaching/research positions and did not even get one interview. I was offered two postdoctoral positions, but with a wife and two children to support, I turned them down and continued applying for tenure track positions. Unlike many of my colleagues, I had an excellent background in electronics and knew I could return to a promising career in that discipline if necessary.

Due to my love for biology and research, I continued looking for a teaching/research position in academia. As graduation approached, I grew more and more concerned. When I was offered a job teaching biology at Rochester Institute of Technology in upstate New York, I accepted without hesitation. I had a heavy teaching load and many different lectures to prepare. Unless I received outside funding, release time for research was unavailable and it looked like my research aspirations would be placed on hold.

With countless job applications still lurking about, at the last minute, I was also offered a summer teaching job at Fort Hays State University in Hays, Kansas. They wanted me to teach a non-majors field-oriented course in biology. Again, I accepted without hesitation. I love outdoor biology and we had lots of bills to pay. The summer job went well, I enjoyed sharing my love for biology with the students, and the students responded positively to my unfettered enthusiasm for all things living.

I still had a large amount of unpublished alligator data from my dissertation research. I knew some of heating and cooling heat flow measurements indicated changes in blood flow, but I did not know how to make the necessary conversions. I ventured into the physics department at Fort Hays and introduced myself to Lou Caplan. I showed him some of my alligator data and explained what I needed. I think he knew how to make the calculations, but knowing it would require a great deal of effort, chose instead to introduce me to his friend, an unsuspecting Stan Robertson. That meeting was one of those rare life-changing events. On a napkin in the cafeteria, Stan started deriving equations he would later use for calculating blood flow in the skin of alligators. Stan and I instantly connected and eventually published some technical papers together. He was able to apply physics to several areas of my research, deepening my understanding of how alligators can heat twice as fast as they cool. We have remained friends for nearly four decades and still occasionally work on papers together.

By summer's end, the biology chairman at Fort Hays seemed impressed with my teaching and student evaluations, and offered me a permanent teaching/research position. Unfortunately, I had already verbally committed to teach at Rochester Institute of Technology and headed northeast. In the farming community I grew up in, a verbal commitment was every bit as binding as a legal document, so I felt I had no choice but to head to New York. Stan and I stayed in touch and in that pre-Internet era, often exchanged letters of a dozen or more hand written pages each week. He was eventually able to calculate the skin blood flow values I needed, but it proved harder than even he expected. Perhaps his friend Lou was wise in passing me on to Stan.

Teaching in upstate New York was an adventure. Being the newest faculty member, I got the least desirable office. The view from my desk was a brick wall ten feet outside my office and during the long, dark winter, it was very cold in my office. On more than one occasion, a cup half full of hot coffee left on my desk during a fifty minute lecture would have ice on the top when I returned. This was with an electric space heater operating under my desk. Besides the severe winter, there were also huge cultural differences. As a private institution, tuition was high and most of the students were from wealthy families. I grew up poor in a farming community and no doubt, they considered me a country bumpkin, a redneck in the most negative sense of the word. To make matters worse, at least once a week I would find myself in a verbal dead end and my only escape was to utter those dreaded words, "Y'all." Without fail, the entire class would break into laughter. Still, it went well and I again had outstanding student evaluations.

Rochester Institute of Technology also had an excellent research grant writing office and helped me secure research money from the King Ranch in

Texas for continuing my alligator research the following summer. Toward the end of my first year, the Dean was so pleased with my performance, that he offered me the Biology Department Chairman's position if I stayed five years. That was a surprising and attractive offer. I saw it as an once-in-a-lifetime opportunity, but I had to turn it down. Unfortunately, due to family matters, I needed to return to Oklahoma.

With a year of teaching experience and excellent letters of recommendation, the job search went well. Soon I was invited to interview for a tenure track teaching position in the biology department at Northeastern Oklahoma State University in the fall of 1977. The Biology Department Chairman was out of town, but the Dean of Science, Dr. Kirk Boatright met me at the airport and we got along splendidly. My research seminar went well and I was offered the position and accepted on the spot. They encouraged research and would provide release time, if I could secure grant money.

Soon we returned to Oklahoma filled with anticipation for a happy future. Our celebration was premature, as it ended poorly just five years later. The second week at my new job, my new boss and chairman of the Biology Department, Everett Grigsby, called me into his office and told me he would never have hired me because he knew I was a Creationist. That is *not* a good way to start a new job, but soon I was immersed in teaching and busy writing research proposals. Within a few months, I helped secure a five million dollar grant to prepare Native American students for graduate and professional schools and it came with release time for research. I was again happy to be teaching biology courses and starting several research projects. It also felt nice returning to my own Oklahoma culture where I knew the quaint Oklahoma dialect. Almost instantly, I seemed to attract a large flock of excellent research students and began research with the heart rate response of wild animals to fear. Soon, I was again using radio telemetry systems I designed and working outdoors with wild animals. The biology department owned a large pontoon boat to which I had access and by spring, I began rabbit, squirrel, chipmunk, and woodchuck studies on a small island in beautiful nearby Tenkiller Lake. My research flourished and life was again good.

My list of scientific publications was growing explosively. With a group of twenty-five or more highly motivated research students each semester, I was working on several different research projects and several led to publications in peer reviewed scientific journals. I was publishing technical papers in three distinct disciplines: reptilian thermoregulation, design and application of sophisticated multichannel radio telemetry systems and my newest topic, the cardiovascular response of wild animals to fear. I also continued publishing Creation and Flood related studies in the *Creation Research Society Quarterly* and began writing children's articles for magazines such as *Highlights* and

Ranger Rick. I also wrote several electronics articles for hobbyist magazines such as ***Popular Electronics*** and other popular hobbyist magazines. My first summer back in Oklahoma, I was given an invitation to study alligators at the Savannah River Ecology Laboratory and accepted. I studied the thermoregulation of a 750-pound alligator and several smaller ones weighing a few hundred pounds. The values Stan and I had previously predicted for large alligators proved valid as more research papers were published. Science is fun!

During the summer of 1978, the British Broadcasting Company filmed my alligator studies for a television documentary, ***A smile for the Crocodile***, describing major crocodilian research around the world. Two portions of the film were devoted to my ongoing alligator research. Due to the worldwide airing of the film, I was invited as keynote speaker to a major biotelemetry conference at Oxford University in England the next year and was able to take along three of my research students. The only other keynote speaker was from NASA and we represented the two extremes in radio telemetry. He had essentially unlimited budgets and each telemetry system he designed cost millions of dollars. I had a limited budget and always feared the telemetered animal might disappear and I would lose the transmitter. My heart rate telemetry transmitters cost less than twenty dollars. We each got to speak for an hour while other attendees were limited to 10 minute presentations. It was the adventure of my life and I made many professional friends. Some have come to Oklahoma to work on various research projects with me.

I took three students with me to the conference and two of them also gave technical papers. One was offered post-doctoral research positions abroad, but she had to decline, as she was only a sophomore. I was also nominated for the prestigious Who's Who in Science and my alligator research was featured in ***Science News***. My woodchuck studies attracted national attention and I was interviewed on the national ***Today*** morning television show. My research was favorably and heavily cited. Sometimes I had three complete pages of citations in the internationally respected ***Science Citation Index***. This is the gold standard for evaluating the quality of a scientist's research. After helping secure the major grant, I had a reduced teaching load and could devote still more time to research and writing. I was truly at the height of my game and enjoying life.

Perhaps it was the arrogance of youth or inexperience, but I never gave my tenure any serious consideration because I knew I was conducting good research and publishing more scientific articles than anyone else in the department. Again, I had outstanding student evaluations and worked with dozens of students that were also conducting publishable research on a variety of topics. This was virtually unheard of at the undergraduate level. At one Oklahoma Academy of Science meeting, my own students completely filled an entire afternoon session. My students and I were soon giving papers at major scientific meetings all over the United States and I had the honor of chairing several sections at these prestigious events. I was doing the work of science and enjoyed it immensely. Tenure was the farthest thing from my mind, but I would soon receive an unexpected reality check.

My last two years at Northeastern, I was selected to teach medical physiology in the newly formed school of optometry. My friend, Stan Robertson, was also teaching in the Physics Department at Northeastern and he was selected to teach optics in the optometry school. This was highly motivating for both of us. The students were bright and we really pushed them. If we professors heard students talking about watching a Sunday football game, we redoubled our efforts. We had to give up television as graduate students and so would they.

I remained active in the area of creation science as well and published several more papers in the *Creation Research Society Quarterly*. I always gave a one-hour lecture in my other science classes about scientific evidences of creation. I did not use Bible references, but simply discussed some of the evidences for creation and the weakness of evidence supporting evolution. Some of my optometry students heard about these lectures and asked me to share that information with them. I felt it improper because the course work in this graduate level professional program was tightly structured. I did not feel I had the freedom or the time to digress for an entire hour of lecture on a topic not related to human physiology. At their unrelenting insistence, I scheduled the lecture for 7:30 on a Saturday morning, thinking no one would show up and I could continue my research. The entire class came and the lively discussion lasted until past noon. The response was insightful, for students had gotten the evolution side of this important debate and seemed hungry to hear an opposing view. One of my own research students became a Christian because of this exposure to the truth.

After five highly successful years at Northeastern, I was once again called into Everett Grigsby's office. I was totally unprepared for what he had to say. He told me my tenure had been denied and I needed to seek employment elsewhere. Certainly, I had options and could have challenged the tenure denial, but I did not want to work where I was not wanted. Shocked and devastated, I returned defeated to the family farm. Unable to earn a living farming, I worked in the oilfield as a roughneck and eventually worked at several other day labor jobs. Due in large part to my loss of professional employment and salary, divorce soon followed ending a thirty-one year marriage. My professional career was over, due to my open acceptance of Biblical Creation. I taught eleven years at a small community college, but there was no opportunity for research and the pay was minimal. Eventually, I became an over the road truck driver for the next twelve years and finally earned enough money for retirement a few years ago.

Sadly, this episode has been repeated thousands of times with other professors. Even worse is the fact hundreds of promising graduate students having been denied access to graduate schools for doubting Saint Darwin. Our battle is real; our Enemy unscrupulous, yet the public remain largely unaware that such blatant religious persecution even exists in the United States. It is widespread in universities, both secular and Christian, where freedom of speech was once celebrated and science is taught as being objective. Neither exists in the increasingly anti-Christian society and the public needs to be aware of this ongoing travesty of justice.

Jerry Bergman, Ph.D.

Perhaps the most thoroughly investigated and best-documented account of religious discrimination is that of Jerry Bergman by Bowling Green State University in Ohio. In many ways, he is a scholar's scholar. He is listed in Marquis Who's Who in the Midwest since 1992, Marquis Who's Who in America since 2000, Marquis Who's Who in Education since the 6th Edition, Marquis Who's Who in Science and Engineering since the 8th edition, Marquis Who's Who in Medicine and Healthcare since 2005, Who's Who in Theology and Science., p. 18-19, Winthrop Publishing Company, Framingham, MA, 1992 and 1996 edition (New York, Continuum 1996 p. 32), and Who's Who in America 47th Ed., New Providence, NJ 07974, Research Centers Directory edited by Maurice Michelle Watkins, Gale Research Company, Book Tower, Michigan, 1990 and in Internationales AdreBregister der Sektenarbeet Munchen 1997-1998. His library is listed in The Directory of Special Libraries and Information Centers (Gale Group, Farmington Hills, MI 2001). His teaching Awards include Who's Who Among America's Teachers for the years 2000 and 2005. He is a member of the prestigious Mensa, the largest, oldest and best-known international society for people who rank in the top two percent of standard intelligence tests and received the 1998 Edgar Langsdorf award for excellence in writing. A fellow of the American Scientific Affiliation since 1983 and currently serving as Biology Editor for the *Creation Research Society Quarterly*, the most respected of the scientific creation scientific journals.

Dr. Bergman is currently an adjunct associate professor at the Medical University of Ohio and teaches biochemistry, biology, chemistry and physics at Northwest State Community college in Ohio. He has taught at the college level for over thirty-five years including seven years at Bowling Green State University, six years at the University of Toledo and twenty years at Northwest State University. He has nine academic degrees with majors in sociology, biology and psychology. He now has over seven hundred publications in a wide variety of scientific and popular journals and has published twenty monographs and books. His work has been translated into twelve languages and he has lectured widely in the United States, Canada and Europe. He is active in the current

ID/evolution controversy and much of his Creation work has been published by *Answers in Genesis*.

In 1978, Bergman was denied tenure from Bowling Green State University where he had been employed since the 1973-74 school year. What is shocking is he received outstanding teaching evaluations by students and faculty. His department chairman and dean recommended him for tenure, yet it was denied. Many more details of his ordeal can be found in his small, but excellent book, **The Criterion** (Bergman, 1984). He has documented that his tenure denial was due to his involvement and publication in the creation movement and for his religious beliefs. He filed a lawsuit with both the Equal Employment Opportunity Commission and the Ohio Civil Rights Commission on the grounds of religious discrimination. Both agencies ruled he was terminated due to lack of peer support and not for his religious beliefs.

He next filed suit against Bowling Green State University in federal district court in 1980, again alleging his right to due process rights had been violated and that he had been denied tenure on the basis of his religious views. That case was dismissed in 1985 and his appeal was turned down in 1987. The court ruled that the reason he was let go was because of ethics, namely that he claimed to have credentials in psychology when, in fact, he "had no psychological credentials." The court rubber stamped the claims of the University. The fact is he originally applied for a test and measurement position at BGSU, but was hired in the psychology area. The reason was, as Dr. Robert Reed in a letter dated Feb. 21, 1973, stated, his "credentials have been ... evaluated favorably by faculty members in the Educational Psychology Area" (emphasis in original). The first classes that he taught at BGSU were in the psychology area, and he taught in that area during the entire seven years he was on the faculty there.

He has a master's degree in social psychology and another in counseling psychology. The minor for his doctorate was in psychology and his dissertation was on an experimental treatment project for second-felony offenders. His graduate course work in psychology includes a total of 131 quarter hours, well over the hours needed for both a masters and a doctorate. Immediately after he left BGSU, he was hired as an associate professor of psychology at Spring Arbor University in Spring Arbor, Michigan. He also has over fifty publications in the area of psychology, including several books. He was a licensed therapist, and trained under Dr. Ricardo Girona and Dr. William Beausay at Arlington Psychological Associates, both licensed psychologists. Until he was licensed, he was registered with the state board of psychology as a psychological assistant. The state law requires that to obtain a license, the candidate must: *"complete a minimum of ninety quarter hours of graduate credit ... including a minimum of thirty quarter hours in: (a) Clinical psychopathology, personality, and abnormal behavior; (b) Evaluation of mental and emotional disorders; (c) Diagnosis of mental and emotional disorders; (d) Methods of prevention, intervention, and treatment of mental and emotional disorders. The individual must complete supervised experience ... supervised by a ... qualified professional approved by the committee"* (Ohio Laws and Rules, 1997, p. 6).

After he met these requirements, he was licensed. His license enabled him to "diagnose and treat mental and emotional disorders" without supervision (Ohio Laws and Rules, 1997, p. 6). The court also ruled that "Dr. Siefert, Dr. Yonker, Dr. Davidson, Dr. Rurke (sic), and Dr. Wiersma... all testified to their negative impressions of plaintiff's work." The fact is, all of these persons admitted in court under oath that they never read any of his publications or could not remember reading them. They did object to a Phi Delta Kappa monograph he authored on the creation evolution controversy, but evidently did not read it. It was clear that their total objections were to Bergman's concerns about the validity of Darwinism.

Caroline Crocker, Ph.D.

The following is mostly from Jerry Bergman (Bergman, 2008) and personal correspondence with Dr. Crocker plus some online sources. Dr. Caroline Crocker did not intend to "start a controversy" when she mentioned intelligent design while teaching her second-year cell-biology course at George Mason University in Fairfax, Virginia. However, many of her colleagues say that the soft-spoken molecular biologist, who received a B.S. in microbiology and virology with honors from the University of Warwick, an MA in medical microbiology from the University of Birmingham, and a Ph.D. in immunopharmacology from the University of Southampton, UK, had "gone too far" (Brumfiel, 2005, p. 1064). Crocker came to doubt Darwinism not because of her religious views, but while doing research for her Ph.D. (Vedantam, 2006).

Recognized as an excellent teacher, a single student accused her of "teaching creationism." In response, numerous students, including an attorney headed for medical school, wrote letters in her defense, noting in their letters that she never "taught creationism" but only discussed her doubts about Darwin. Nor did she ever mention religion or the Bible in her classes and the students did not know what religion she was or where she stood on the Darwin/ID issue.

Dr. Crocker has now been "barred by her department from teaching both evolution and intelligent design. 'It's an infringement of academic freedom,' she says. She appealed the case to a grievance committee (Brumfiel, 2005, p. 1064). Because she did not prevail in her grievance, criticizing Darwin cost her a promising career: "Her lectures drew criticism from some and praise from others, notably, she says, her Muslim students seemed to like it. She maintains that the talks help students to think independently about ideas such as evolution" and intelligent design. The former Oxford University scientist with twenty-nine major scientific publications explains that her goal is to teach students to think for themselves. Many of her students were very supportive of her teaching.

Dr. Crocker's students did not want their names used in the media for fear of retaliation for their beliefs in support of ID (Vedantam, 2006). Brumfiel concludes by noting whether, and in what form, Dr. Crocker's discussions about ID will continue is up to the faculty members and administrators. "The university doesn't have a policy or a rule on whether certain topics should be discussed," says Daniele Struppa, a mathematician and dean of the College of Arts and Sciences at George Mason University. But, he adds, he questions whether a concept with theological underpinnings really belongs in a science course. Darwinists are divided over whether intelligent design deserves a classroom airing. Forrest says that she believes, "This is not a question of academic freedom, this is a question of professional competence." But Eugenie Scott, director of the National Center for Science Education in Oakland, California, which vehemently opposes teaching intelligent design in high schools, takes a different view. She thinks such discussions are more acceptable in a college environment, but believes it must be made clear to students that intelligent design is theology, not science. Crocker is one of a handful of professors nationwide who are introducing intelligent design into college-level teaching. Some, like Crocker, try to work the idea into their biology classes, but increasingly, intelligent-design advocates are teaching their material outside the science curriculum in special seminars and one-time courses, says Barbara Forrest, a philosopher at Southeastern Louisiana University (Brumfiel, 2005, p. 1064).

Progress has been made because teaching intelligent design is now met with a mixed response from faculty members and administrators on campus . Michael Behe, an intelligent-design advocate and biochemist at Lehigh University in Bethlehem, Pennsylvania, teaches an elective first-year seminar on 'popular arguments on evolution.' "The majority of my colleagues disagree with me. But my chairman supports my right to have my own views and argue them in a public setting" (Brumfiel, 2005, p. 1064).

Caroline Crocker was self-employed in the Washington DC area as a senior science writer, consultant, and private tutor. She just finished her first book, *Science Censored*, on her experiences as a full time university lecturer who strove to present Darwinian evolution from an intellectually honest viewpoint. Dr. Crocker taught various biology courses for five years at George Mason University (GMU) and Northern Virginia Community College. From all indications, she was an excellent teacher and was certainly knowledgeable in here area of expertise. While at GMU, she was awarded three grants, including one from the Center for Teaching Excellence, commendations for high student ratings, and wrote a cell biology workbook. Dr. Crocker did her post-doctoral studies in fluorescence resonance energy transfer analysis of interactions between proteins of the T-cell receptor signal transduction pathway at the Uniformed Services University.

While doing a Ph.D. in immunopharmacology (The Modulation of Phosphodiesterase Activity in Human T-Lymphocytes) as an external student at Southampton University, U.K., she worked at Creighton University as a research associate, bringing in numerous grants, conducting basic immunology research, and publishing extensively. She received a M.Sc. from Birmingham University, a B.Sc. from Warwick University, and an A.A. from Des Moines Area Community College, having graduated from high school at the age of sixteen. Dr. Crocker is married to Richard and they have four grown children.

She has recently accepted a position in California working with the Idea Club program which encourages open discussion about creation and evolution in high school and university classrooms. For additional information visit: www.ideacenter.org/clubs. For additional information about this courageous young woman, read an in depth account online at: www.washingtonpost.com/wpdyn/content/article/2006/02/03/AR2006020300822 _2.html

Guillermo Gonzalez, Ph.D. is perhaps the best known and most publicized recent victim of tenure denial. He was denied tenure at Iowa State University in spite of his publication of sixty-eight peer-reviewed scientific articles. This was four times what the department suggests as required for a minimum standard of excellence. In addition, his scientific articles have the highest citation rate among all of the astronomers in his department. This is the academic standard used to evaluate the quality of papers published by professors. It is not enough to write and publish peer reviewed research articles, but they must be read and cited by others working in that particular area. His work has been featured in such prestigious magazines as *Science*, *Nature* and *Scientific American*, which did a cover story about his cutting-edge research.

Everyone agrees he did not teach design in his classes, but was denied tenure because of the popular astronomy book he wrote, *The Privileged Planet: How our place in the cosmos is designed for discovery*. The book was later made in to the popular movie, *The Privileged Planet*. Dr. Gonzalez was targeted by atheist professors, including one in his own department in 2005. They drafted a petition against intelligent design in the science curriculum and collected 120 signatures from faculty. Hector Avalos, an outspoken atheist, told the *Ames Tribune* the petition was motivated by growing attention given to Gonzalez's work and concerns the university would be seen as an "intelligent design school." Avalos, a religion professor at Iowa, believes the Bible is worse than Hitler's *Mein Kampf*. Geoffroy, one of Gonzalez's chief persecutors, was promoted to full professor. Incredibly, Iowa State University's President Geoffroy denied

tenure to Gonzalez while approving 91 percent of the others applying for tenure the same year. The next day, President Geoffroy admitted the only reason for Gonzalez's tenure denial was his support for intelligent design. The university faculty issued a statement saying claims for intelligent design *"are premised on the arbitrary selection of features claimed to be engineered by a designer, unverifiable conclusions about the wishes and desires of that designer and an abandonment by science of methodological naturalism. Whether one believes in a creator or not, views regarding a supernatural creator are, by their very nature, claims of religious faith, and so not within the scope or abilities of science. We, therefore, urge all faculty members to uphold the integrity of our university of 'science and technology,' convey to students and the general public the importance of methodological naturalism in science, and reject efforts to portray intelligent design as science."* For his part, Gonzalez defended the attack by noting that he had put forth a design argument that was scientifically rigorous, testable and falsifiable, but whatever its merits, he had not taught the argument in the classroom. Additional information can be found online at the following websites: WorldNetDaily: Intelligent design scientist denied tenure, www.wnd.com/news/article.asp?ARTICLE_ID=55667 and
CLA - Christian Law Association Online - Has Modern Evolutionary "Science" Become Unscientific?

Nathaniel Abraham, Ph.D., is a biologist, scholar, and research scientist specializing in zebra fish, developmental biology, and programmed cell death. Much of the following information was obtained online from www.Christianlaw.org. He was fired from Woods Hole Oceanographic Institution, a highly esteemed research facility based in New England in March of 2004. The reason was that Dr. Abraham believes in Creation and does not accept evolution as a scientific fact. With his termination, Dr. Abraham became one of many scientists and academics who have dedicated years of their lives to the pursuit of scientific knowledge; but because they are unwilling to spout the "party line" with respect to evolution, they are being forced out of their field.

Dr. Abraham came to the United States from India to earn his Ph.D. in biology at St. John's University in New York. While at St. John's, he became an expert in zebra fish. He did a groundbreaking doctoral dissertation in his area of research science. Dr. Abraham was well respected at St. John's, and it was no secret that he was a strong Christian who believed the Bible and believed in Creation. Many of his colleagues there enjoyed asking him questions about issues of science and religion. In the first sentence of the acknowledgements in his doctoral thesis, Dr. Abraham thanks his Lord and Savior Jesus Christ.

While Dr. Abraham was completing his Ph.D. program at St. John's, he began to look for a postdoctoral position where he could continue his research.

He married, brought his wife to America, and they were expecting their first child. Life was looking very good. He applied for a research position at Woods Hole after seeing the opportunity listed on the Internet. Scientists at Woods Hole were interested in expanding their toxicology laboratory to include programmed cell death research using the zebra fish benchmark model, which was Dr. Abraham's specialty. In fact, Dr. Abraham was so valuable to them in this endeavor, that he assisted them for several months in designing and building their lab before he was able to join them fulltime as a researcher. Dr. Abraham contributed greatly to the design of the new facility and his superiors were pleased with his work. The relationship between Dr. Abraham and Woods Hole was mutually beneficial and productive shortly after Dr. Abraham began working fulltime at Woods Hole. In a passing conversation with his supervisor over lunch, Dr. Abraham happened to mention that he believed in Creation. His supervisor expressed concern that if Dr. Abraham believed in Creation, he was not qualified to do the job he had been hired to do.

Dr. Abraham told his superiors in writing that he was willing to analyze his research using evolutionary concepts if that was warranted, but he was not willing to personally renounce his belief in Creation or accept evolution as a scientific fact. Dr. Abraham pointed out that his belief in Creation was irrelevant to the research he was doing. After all, he had developed his research specialty and expertise at St. John's University, where his faith had not been an issue. Dr. Abraham believed a scientist should merely do the experiments and follow the evidence wherever it led. The "established" scientists at Woods Hole were not convinced. They wanted Dr. Abraham to share their wholehearted acceptance of evolution as a scientific fact, or they felt he was not "fit" to work in the field of science, at least not at their prestigious laboratory. They finally told Dr. Abraham he would have to leave if he did not accept evolution as fact.

Dr. Abraham's decision to hold to his Biblical faith and to his belief in Creation was a very costly decision for him. His wife was about to have their baby. Without a job, Dr. Abraham would lose his visa, which would soon expire. Woods Hole was not willing to even permit Dr. Abraham to complete the first year of his potential three-year employment. As a result, Dr. Abraham was forced to send his expectant wife home to India, and he missed the birth of his first child. Attorneys for the Christian Law Association are now representing Dr. Abraham in a lawsuit against Woods Hole to test whether America's courts will permit this sort of blatant religious discrimination against scientists in America.

In summary, science is no longer a search for objective truth. Scientists do not have the freedom to follow the evidence from their discoveries, if it leads to design or to the Creator. This is tragic and it means science as we have known it is on the verge of collapse. Science must again be truly objective. If the scientific evidence leads to design, so be it. Indeed, it must be so. ***The heavens*** *declare the glory of God; the skies proclaim the work of his hands. Day after day they pour forth speech; night after night they display knowledge. There is no speech or language where their voice is not heard.* (Ps 19:1-3)

Chapter 15
Does It Really Matter?

Salvation is found in no one else, for there is no other name under heaven given to men by which we must be saved. (Acts 4:12, NIV)

Few topics are more important than understanding our origin. Our personal worldview rests on the foundation of our acceptance of origins. Why am I here? Where did I come from? What is the purpose of life? Is there life after death? Is God real? Are there absolutes in the areas of morality and ethics? If so, what are they and is there such a thing as right and wrong? These and related questions have plagued mankind throughout history. The answers to these fundamental questions influence our decisions and have eternal consequences. Society and secular education teaches man is nothing more than an animal resulting from millions of years of undirected, mindless evolution and genetic mistakes. Students today are taught ethics and moral values of right and wrong are nothing more than myth or superstitions of the uneducated. They are told man created God, not the other way around. Life after death is superstition lacking scientific support. Human life is nothing more than matter in motion. This worldview permeates our culture and is taught in our public schools and universities. Let's consider the alternative worldview and why the topic of origins is of great importance.

What if the Bible really is true? What if modern science is wrong about origins and evolution is false? What if God created the universe and everything in it, including man? What if this Creator-God loves us and has a divine purpose and plan for our lives? Nothing could be of greater importance. Suddenly life has meaning. Heaven, hell and final judgment become realities. The question of origins is not only a key issue in understanding science, but is also necessary to understand ourselves. Our understanding of origins influences our lives and our daily decisions and behavior. Let's consider one of the most famous wagers in history, for it addresses this important issue.

Blaise Pascal was a seventeenth century child prodigy and important French mathematician, physicist and religious philosopher. Pascal's wager posits it is far better to bet on the possibility that God exists than to not believe. The reason is obvious. Little is lost and everything gained if one believes and God is real. The risks are high and all is lost if one fails to believe, but God turns out to be real. In other words, the value of believing is far greater and the risks less than of not believing. For this reason alone, many believe. Fearing God can be a good and powerful thing.

The Apostle Paul saw it as good. Continue to work out your salvation with fear and trembling, for it is God who works in you to will and to act according to his good purpose. (Phil 2:12b-13) He again confirmed this in Ephesians. Bondservants, be obedient to those who are your masters according to the flesh, with fear and trembling, in sincerity of heart, as to Christ; not with eye service, as men-pleasers, but as bondservants of Christ, doing the will of God from the heart, (Eph 6:5-6, NKJ) Certainly the God of the Bible is love, but He can also be vengeful. King David saw this graphically displayed after he wrongly took the infamous census of his fighting troops. He witnessed seventy thousand of his own people killed in a single day by a vengeful God. (Read the full account in 2 Samuel 24.)

An excellent example of this teaching is found in the classic sermon given by Jonathan Edwards in Enfield Connecticut on July 8, 1741. Many consider him the greatest theologian America has produced. The sermon was titled simply, *Sinners in the Hands of an Angry God.* It is still studied today as the epitome of hell fire preaching. His sermon text was simply: ***It is mine to avenge; I will repay. In due time their foot will slip; their day of disaster is near and their doom rushes upon them.*** (Deut 32:35) With heart wrenching detail, he graphically illustrates that for the boundless Grace of God we would all burn in a fiery hell as we rightly deserve. (The complete sermon can be found online at: www.cced.org/cced/edwards/sermons.sinners.html) Many are slow to admit that most of us become Christians more out of fear of an angry God than we are attracted to the love of God. Certainly both are important.

Time and time again history, archeology and science have verified the truth of the Holy Scripture. As we have seen, there are over one thousand five hundred Bible passages proclaiming God as Creator and Sustainer of our world. Any topic repeated this many times is clearly of great importance. The reason there are so many Bible verses describing God as Creator is not to provide scientific insight into how it was accomplished. Instead, the primary purpose of these passages is so we might look at nature and see God. In doing so, we are to lift our hands and hearts in worship and adoration of the Creator as was done my the majority of scientists prior to Darwin. We must never worship the creation as many today do through godless evolution. Again, it is obvious to all, ***The heavens declare the glory of God; the skies proclaim the work of his hands. Day after day they pour forth speech; night after night they display knowledge. There is no speech or language where their voice is not heard.*** (Ps 19:1-3, NIV) It is for this reason, ***The fool says in his heart, "There is no God." They are*** *corrupt, their deeds are vile; there is no one who does good.* (Ps 14:1) Man is without excuse. God is self-evident from the wonders of His Creation.

There are other reasons for the many references to God as Creator. Through His Creation, we learn the power and character of God. In Creation, we also see evidence of the Son of God and our Savior. From Genesis we find *Then God said, "Let us make man in our image, in our likeness, and let them rule over the fish of the sea and the birds of the air, over the livestock, over all the earth, and over all the creatures that move along the ground." So God created man in his own image, in the image of God he created him; male and female he created them.* (Gen 1:26-27, NIV) The "us" in this passage refers to God the Father, Jesus the Son and the Holy Spirit. Each person of the Godhead played an active role in the Creation process. *Now the earth was formless and empty, darkness was over the surface of the deep, and the Spirit of God was hovering over the waters.* (Gen 1:2) Jesus Christ Himself was involved in the original creation.

His involvement in the original Creation is clearly confirmed in many passages of scripture. Consider, for example, the Gospel of John. It begins with this important truth. *In the beginning was the Word, and the Word was with God, and the Word was God. He was with God in the beginning. Through him all things were made; without him nothing was made that has been made. In him was life, and that life was the light of men.* (John 1:1-4) We are to worship Jesus the Son of God, not only as our Savior, but also as our Creator. His being active in the original Creation provides undeniable credibility that He is indeed equal to God the Father and worthy or our adoration, praise and worship. This credibility is central to our trusting Jesus for salvation. The most important question in all of life is, "What will you do with Jesus?" Our salvation for all eternity depends on our response to this most important question. Our acceptance or rejection of God as Creator influences that decision. God's word clearly states: *And as it is appointed for men to die once, but after this the judgment, so Christ was offered once to bear the sins of many. To those who eagerly wait for Him He will appear a second time, apart from sin, for salvation.* (Heb 9:27-28, NKJ) Only if Jesus is God, can He forgive our sins. His role in Creation demonstrates that He is the all wise and all powerful God. He alone is worthy of our worship and praise. Again, we must understand and be able to defend this God of Creation. The stakes for a lost and dying world could not be higher. Are you prepared? The risks are also great. Will you witness no matter what it costs?

What Of Other Religions?

In today's popular culture of tolerance and relativism, it is not politically correct to consider Christianity better than the other religions of the world.

Society teaches, "Choose the path that suits you." This is a lie from the pit of hell. The Bible is not politically correct. It does matter where you place your faith and in whom you believe. Only the Living God, the Creator of the Bible has provided the way to eternal life. Tolerance of other gods or faiths is unacceptable and contrary to the clear teaching of the Bible. Consider again the opening Bible verse for this chapter. ***Salvation is found in no one else, for there is no other name under heaven given to men by which we must be saved.*** (Acts 4:12) Not all religions are the same. Any religion that rejects Jesus Christ as Lord leads to eternal damnation and despair. The Creator of the universe will accept no other than worship of His Son, Jesus Christ. Only through Him can we gain salvation for all eternity. Yes, there are other paths, but they all lead to death, judgment and eternal punishment. There is no other way. Remember these life-giving words of Jesus: ***"Enter through the narrow gate. For wide is the gate and broad is the road that leads to destruction, and many enter through it. But small is the gate and narrow the road that leads to life, and only a few find it.*** (Matt 7:13-14) ***Jesus answered, "I am the way and the truth and the life. No one comes to the Father except through me.*** (John 14:6)

All other religions are false and worship false gods. There is no more important topic worthy of consideration. What will *you* do with Jesus the Christ? This is the most important question of the ages and is very much dependent upon your acceptance or rejection of the truth of the Creation account in the Bible. ***Have nothing to do with godless myths and old wives' tales; rather, train yourself to be godly*** (1 Tim 4:7). As in other areas, if God is truly the Creator and the Bible can be trusted, the implications for Christians are tremendous; everything changes. A few areas of considerable importance will be discussed.

Praise and Adoration

The hundreds of Bible passages proclaiming God as Creator are there for one simple reason, so that as we see the breathtaking beauty and complexity of nature we will lift our hearts and hands in honor of Him who made it all. We will truly be able to praise His name with the same honor and enthusiasm as King David long ago. ***I praise you because I am fearfully and wonderfully made; your works are wonderful, I know that full well. My frame was not hidden from you when I was made in the secret place. When I was woven together in the depths of the earth, your eyes saw my unformed body. All the days ordained for me were written in your book before one of them came to be. How precious to me are your thoughts, O God! How vast is the sum of them! Were I to count them, they would outnumber the grains of sand.*** (Ps 139:14-18) He alone is to be rightfully worshiped as Creator of all things. This is the opposite of what is seen on television and taught in our public schools and universities where God is robbed of His rightful place as Creator and the honor and worship are given instead to the laws of nature…which God also made.

Personal Bible Study

Over and over I have seen Christian students lose faith in the Bible when exposed to godless evolution. It breaks my heart each time. Their confidence in the truth of the sacred book vanishes, and with it, their motivation to reading and contemplating God's instruction manual. Although the root cause was exposure to evolution and questions about Creation, soon this shadow of doubt encompasses all scripture. For many, such exposure stops Christian growth and soul winning. If, instead, the Bible is seen as trustworthy personal time studying God's Truth will again become important. The wisdom of the Bible will be sought. Lives will be strengthened and blessed. A hunger for reaching others with the truth of God's Word will increase and with it witnessing and wining souls. Nothing is of greater importance than confidence in the Word of the Living God. For it has the power to change hearts and lives. We have God's promise on this. *So is my word that goes out from my mouth: It will not return to me empty, but will accomplish what I desire and achieve the purpose for which I sent it.* (Isa 55:11) It has the power to protect from sin. King David asked this question and clearly gives the answer. *How can a young man keep his way pure? By living according to your word. I seek you with all my heart; do not let me stray from your commands. I have hidden your word in my heart that I might not sin against you. Praise be to you, O LORD; teach me your decrees.* (Ps 119:9-12) Let's once again have confidence in the power of those powerful pages at our fingertips or on our computer. Let our nation return to having the confidence in the Bible it had when our great country was founded.

Praise God as Author

We have long worshiped Jesus as Author of our salvation and this is as it should be. Let us fix our eyes on Jesus, the author and perfecter of our faith, who for the joy set before him endured the cross, scorning its shame, and sat down at the right hand of the throne of God. (Heb 12:2, NIV) There is more. With the recent elucidation of the complete human genome and the explosion of our knowledge about the workings of DNA, we can also worship God as Author of every kind of living thing. When he created the animals and brought them to Adam to name, He also commanded each kind to "reproduce after its kind." That alone should prove evolution can't occur. In order to reproduce after its kind, God used DNA to write out the formula or genetic description of every living thing. In a very real sense we can worship God as Author of our salvation and as Author of every living thing. For this, let's praise Him. He is the God who even knows the number of hair in our head. Are not two sparrows sold for a penny?

Yet not one of them will fall to the ground apart from the will of your Father. And even the very hairs of your head are all numbered. So don't be afraid; you are worth more than many sparrows. (Matt 10:29-31) How can anyone read that passage and not feel overwhelmed with love and adoration for the God of Creation…who loves each of us with such an unfathomable love! Let us praise Him for He alone is worthy of our endless praise.

End abortion

The most tragic application of evolution is the senseless slaughter of millions of unborn babies. Evolution is at the foundation and justification of abortion for it teaches humans are but animals and have no intrinsic value or purpose. Many years ago, a beloved pastor of mine said future generations will look back in disbelief at our killing of the most innocent and helpless of all humans. I agree. The Nazi death camps of the last century were nothing when compared to the killing of the unborn in this country. America will be punished for this cruelty. Perhaps it has started even now.

As Christians we *must* put an end to this tragic loss of human life and potential.

Abortion is wrong…humans have value and purpose. There was an old cartoon in a Christian magazine that illustrates this point. In the first panel, an old man was pouring his heart out to God asking Him to send someone with the cure for AIDS. In the next panel, the man is shocked to hear a voice from Heaven that proclaims, "I have already sent him." The old man asks where he is, for AIDS continues to decimate the population? God replies, "I sent him, but he was aborted."

Stop Euthanasia

Likewise, Christians must ban euthanasia. Now the terminally ill can be murdered legally in some states and the practice is spreading. As with abortion, euthanasia has its roots deep in Darwinian soil for evolution teaches life has no value and humans are but animals without a soul. There is talk of killing the elderly to avoid soaring medical costs. Will we allow a repeat of Nazi Germany within our own borders? It seems to be blowing in the wind. Christians must let their voices be heard for the time is short.

Education

By all accounts, our public schools are failing. This has increased with the removal of prayer, the Bible and Ten Commandments. Again, Christians must become involved. Many feel the only solution is Christian schools or homeschooling. In test after test and countless surveys, home schooled children learn more and perform better than their counterparts attending public schools.

When I was teaching at a community college all incoming students were tested. Fully 85% of public school graduates were functionally literate and required up to two full years of remedial course before they could begin our watered down "college" courses. There is another advantage. The truth of the Bible and Creation can be freely taught at home, something outlawed in our government schools. The future generation is at stake. Let's get involved.

God's word can be trusted regarding Creation and in all other areas to which it speaks. Let's again allow the Living Word to permeate and give meaning and direction to our lives and the lives of our children. Let's never apologize for teachings and morality found throughout the Bible. We must have absolute confidence in it's teachings for our society desperately needs this direction. Help our students recognize the lies of godless evolution in textbooks and classes. Time is short; our enemy is ruthless. Parents must again, ***Train a child in the way he should go, and when he is old he will not turn from it.*** (Prov 22:6, NIV) Let's never forget, ***Salvation is found in no one else, for there is no other name under heaven given to men by which we must be saved.*** (Acts 4:12, NIV)

Yes, I am coming soon. Amen. Come, Lord Jesus. (Rev 22:20b, NIV)

THE END

References Cited

Allen, L., 2002. Unlocking the Mystery of Life: The Scientific Case for Intelligent Design. (La Habra, CA: Illustra Media, 2002). (film)

Alters, Saundra, 2007 Biology: Understanding Life. Published by John Wiley and Sons. ISBN-10: 0470240601

Ankerberg, John and John Weldon, 1998. Darwin's Leap of Faith: Exposing the False Religion of Evolution. Harvest House Publishers. ISBN-10: 1565076575

Austin, S. 1994. Grand Canyon: Monument to Catastrophe, ICR, p 145.

Ayala, F. and James W. Valentine. 1979. *Evolving*. California, Benjamin Cummings Publishers.

Ayala, F. 2007 "Darwin's greatest discovery: Design without designer," *Proceedings of the National Academy of Sciences* published online before print, May 9, 2007.
Barzun, J. 1959 *Darwin, Marz, Wagner*, 2nd ed, Garden City, NY: Doubleday & Co.

Behe, Michael J.; Colanter, Eddie N.; Gage, Logan; Johnson, Phillip. 2008. Intelligent Design 101: Leading Experts Explain the Key Issues. Published by Kregel Publications, ISBN-10: 0825427819.

Bergman, Jerry. 1984. *The Criterion: Religious Discrimination In America*. Onesimus Publishing. 1984.

Bergman, Jerry. 2008. Slaughter of the Dissidents. Leafcutter Press. ISBN-10: 0981873405.

Brumfiel, Geoff. 2005 Cast out from class: Intelligent design: Who has designs on your students' minds?" *Nature,* April 28, 2005. 434:1062-1065. See page 1064

Budd, Graham E. and Maximilian J. Telford. 2005. "Evolution: Along came a sea spider," *Nature,* vol. 437, Oct. 20, 2005, p.1099.

Buell, Jon. 1994. Darwinism: Sciences or Phylosophy? Foundation for Thought and Ethics, Dallas Christian Leadership and C. S. Lewis Fellowship.

Bull, J. J. and R. C. Vogt. 1979. Temperature-Dependent Sex Determination in Turtles. Science 206:1186-1188.

Cameron & du Toit. 2007. "Winning by a neck: tall giraffes avoid competing with shorter browsers." *The American Naturalist*: January 2007.

Colbert, Edwin H, Michael Morales and Eli C. Minkohh. 2001. *Colbert's Evolution of the Vertebrates*, Wiley-Liss, 5th edition. ISBN-10: 0471384615.

Crick, Francis. 1990. What Mad Pursuit. Published by Basic Books. ISBN-10: 0465091385

Crick, Francis. 1982. Life Itself: It's Origin and Nature. Simon and Schuster Adult Publishing Group. ISBN-10: 0671255630.

Crowther, J. G. 1935. British Scientists of the 19th Century. London, K. Paul, Trench, Trubner and Co. pp. 138-140.

Darlington, P. "The Origin of Darwinism" *Scientific American,* 1959. **200**(5): 61-62.

Darwin, Charles. 1968. The Variations of Animals and Plants under Domestication published by John Murray publishers of London. January 30, 1868.

Davis, P., and D. H. Kenyon, 1989. *Of Pandas and People: The Central Question of Biological Origins* (Richardson, TX: Foundation for Thought and Ethics). ISBN 0914513400

Dawkins, R., 1987. *The Blind Watchmaker*

Dawkins, R. 2008. The Blind Watchmaker

Denyse O'Leary, 2004. By Design Or By Chance? The Growing Controversy On The Origins Of Life In The Universe. Published by Castle Quay. ISBN-10: 1894860039.
Dembski, W.A. ed., 1998. *Mere Creation: Science, Faith & Intelligent Design* (Downers Grove, IL: InterVarsity Press). ISBN 0830815155

Denton, M., 1985. *Evolution: A Theory in Crisis* (Bethesda, MD: Adler & Adler,), 341. ISBN: 0917561058

Desmond and Moore. 1991 *Darwin*, Michael Joseph, London, Penguin Books.

Dietz, R. 1983. In defense of drift. The Sciences 23:22-26.

Donoghue, P. 2007. Embryonic Identity Crisis, *Nature*, p. 155

Eldridge, Niles, 2001, The Triumph of Evolution and the Failure of Creationism, Published by Holt Paperbacks, ISBN-10: 0805071474.

Eiseley, L. "Alfred Russel Wallace," *Scientific American*, 1959. V. 200, p. 70.

Eschberger. Grasses and grazers. *Suite 101.* July 7, 2000.

Fisher, A. 2003. *Grolier Multimedia Encyclopedia*, fossil section.

Gonzalez, Guillermo, Richard Sternberg and Caroline Crocker. The facts about the "expelled" scientists in expelled, an article from*: Skeptic.* Altadena, CA

Gould, Stephen J and Niles Eldredge. 1977. Abstract to 'Punctuated Equilibria: the tempo and mode of evolution reconsidered' *Paleobiology*, v. 3, 1977, p. 115.

Gould, S.J. 1980. The Episodic Nature of Evolutionary Change *The Panda's Thumb* W.W. Norton & Co. 1980.

Gould, Steven Jay, 2007. Richness of life: The Essential Stephen Jay Gould, published by W. W. Norton and Company, ISBN-10: 0393064980

Grassé, Pierre-P. *Evolution of Living Organisms* (New York: Acad. Press, 1977), p. 130.

Harris, C. *Evolution – Genesis & Revelations*, State U. of N.Y. Press, Albany, 1981, p. 140-144.

Harrison, R. K. 1999. *Introduction to the Old Testament.* Peabody, MA: Hendrickson Publishers, 508. (As cited from Josh McDowell's book, *The New Evidence that Demands a Verdict*, San Bernardino, CA: Here's Life Publishers, 1999: 409)

Hecht, J. 2005. Large mammals once dined on dinosaurs. *New Scientist.com*, January 15, 2005.

Hoaglan, H. 1964. Science and the new humanism. Science 143:113.

Hoyle, F., 1982. "Evolution from space" (Omni Lecture) (London: Royal Institution, January 12, 1982); also, F. Hoyle, and N. C. Wickramasinghe, *Evolution from Space: A Theory of Cosmic Creationism* (New York: Simon and Schuster, 1982). ISBN 067145031X.

Hsu, K. and J. McKenzie. 1986. Rare events in geology discussed at meeting. Geotimes 31:11-12.

Huxley, J. 1959. Associated Press. Address at Darwinian Centennial Convocation, Chicago University, November 27, 1959. (See Issues in Evolution.)

Jacobsen, N. K. 1979. Alarm bradycardia in white-tailed deer fawns, *Odocoileus virginianus . J. of Mammal.* 60:343-349.

Jammer, Max. 2002 *Einstein and Religion: Physics and Theology*, Princeton University Press. ISBN-10: 069110297X

Johnson, P. E., *Darwin On Trial* (Washington, DC: Regnery Gateway, 1991), 144. ISBN 0895265354

Linnaeus, C. 1730. Prelude to the Nuptials of Plants.

Lovtrup, S. 1987 *Darwinism: the refutation of a myth*, N.Y. Croom Helm.

Lewontin, Richard 1997. The New York Review, p. 31.

Larson, Edward and Larry. 1999. Scientists and Religion in America. Scientific American, 1999.

Light, *Funk & Wagnalls 1995 Science Yearbook,* p. 54.
Maisey, J. G. 1996 of the American Museum of Natural History.

Miller, Kenneth R. and Joseph S. Levine. 2007. Prentice Hall Biology, published by Pearson Prentice Hall. ISBN-10: 0132013495

Morio Beauregard and Denyse O'Leary, 2007. The Spiritual Brain: A Neuroscientist's Case for the Existence of the Soul. Published by HarperOne. ISBN-10: 0061625981.

Morris, Henry M. 1982, Men of Science-Men of God. Master Books, Inc. 107 pages. ISBN: 0-89051

Neufeld, Michael. 2008. Von Braun: Dreamer of space, Engineer of War. Vintage, 624 pages. ISBN-10: 0307389375.

Norris, S. 2006. Many dinosaur fossils could have soft tissue inside, *National Geographic News*, February 22, 2006.

Ohmoto, Hiroshi, YUMIKO Watanable, Hiroaki Ikemi, Simon R. Poulson, Bruce E. Taylor. 2006. Sulphur isotope evidence for an oxic Archaean atmosphere. Nature 442:908-911.

Palmer, Trevor. 1999. Controversy—Catastrophism and Evolution. Published by Springer, 468 pages. ISBN-10: 0306457512.

Robertson, Stanley L. and Darryl J. Leiter, 2002. Evidence for Intrinsic Magnetic Moments in Galactic Black Hole Candidates. The Astropysical Journal 565:447.

Robertson, Stanley L. and Darryl J. Leiter. 2003. On Intrinsic Magnetic Moments in Black Hole Candidates. The Astrophysical Journal Letters 596:L203.

Robertson, Stanley L and Darryl J. Leiter. 2004. On the Origin of the Universal Radio-X-Ray Luminosity Correlation in Black Hole Candidates. Monthly Notices of the Royal Astronomical Society 350:1391.

Robertson, Stanley L and Darryl J. Leiter. 2006. The Magnetospheric Eternally Collapsing Object (MECO) Model of Galactic Black Hole Candidates and the Active Galactic Nuclei. Nova Science Publishers. Chapter: New Developments in Black Holes Research.

Schild, Rudolph E., Darryl J. Leiter and Stanley L. Robertson. 2008. Direct MicroLensing-Reverberation Observations on the Intrinsic Magnetic Structur of

Active Galactic Nuclei in Different Spectral States: a Tale of Two Quasars. Astronomical Journal 135:947.

Schild, Rudolph E., Darryl J. Leiter and Stanley L. Robertson. 2006. Observations Supporting the Existence of an Intrinsic Magnetic Moment Inside the Central Compact Object Within the Quasar Q0957+561. Astronomical Journal 132:420.

Schwabe, C., 1994. As quoted in "Hox (homeobox) Genes — Evolution's Savior?" by Don Batten, answersingenesis.org/docs/4205.asp

Schweitzer, M. et al., *Science*, v. 307, p. 1952-1955.

Scott, Eugenie C., 1987. Creationism lives, Nature, correspondence 329:282

Shea, J. 1982. Twelve fallacies of uniformitarianism. Geology 10:455-460.

Short, Philip, 2006. Pol Pot: Anatomy of a Nightmare, published by Holt Paperbacks, ISBN-10: 0805080066

Smith, E. N. 2006. Passive Fear: Alternative to Fight or Flight, published by iUniverse. ISBN: 0-595-67665-0.

Smith, E, N. 2007. Endothermic Skunk Cabbage. *Creation Research Society Quarterly*. 44(2):153-155.

Smith, E. N. and J. Swallow. 2010. American Buffalo and Native Americans. The Perfect Circle Publishing Company. ISBN:

Stahl, B. 1974. Vertebrate History: Problems in Evolution, Dover.

Straus, Jr. W. L. 1947. Book review. Dating the Past: An Introduction to Geochronology by Frederick E. Zeuner. The Quarterly Review of Biology. December 1947, Vol. 22, no. 4

Strobel, Lee. 2005. The Case for a Creator: a journalist investigates Scientific Evidence that points toward God. Zondervan, 352 pages. ISBN-10: 0310240506.

Swift, D. 2002. *Evolution under the Microscope*, Leighton Academic Press, 2002.

Thayer, H. S. *Principia*, Book III; cited in; Newton's Philosophy of Nature: Selections from his writings, ed. H.S. Thayer, Hafner Library of Classics, NY,

1953, p. 42. A Short Scheme of the True Religion, manuscript quoted in Memoirs of the Life, Writings and Discoveries of Sir Isaac Newton by Sir David Brewster, Edinburgh, 1850; cited in Newton's Philosophy of Nature, cited in Thayer, 1953, p. 65.

Thaxton, C. B., W. L. Bradley, and R. L. Olsen, 1984. *The Mystery of Life's Origin.* (Dallas, TX: Lewis and Stanley), 210-211. ISBN 0802224466
Trefil, James, 1996, The Edge of the Unknown: 101 Things You Don't Know about Science - and No One Else Does, Either. Published by Houghton Mifflin, ISBN-10: 0395728622.

Trombly, J. 1995. Cast a Cold Light, *Funk & Wagnalls 1995 Science Yearbook,* p. 54.

Trotsky, Leon. 2007. My Life: an attempt at an Autobiography. Dover Publications, ISBN-10: 0486456099

Tumarkin, Nina. 1997. Lenin Lives: The Lenin Cult in Soviet Russia. Published by Harvard University Press. ISBN-10: 0674524314

Unger, K. *Science* NOW Daily News, Dec. 2, 2005, citing *Journal of Experimental Biology.*

Vedantam, Shanker. "Eden and Evolution." *Washington Post.* February 5, 2006. p. W8.

Wiedersheim, R. 1895. Vestigial organ list (1893) The Structure of Man: An Index to His Past History. Second Edition. Translated by H. and M. Bernard. London: Macmillan and Co. 1895.

Yeoman, B. 2006. Schweitzer's Dangerous Discovery, *Discover*, April, 2006, p.37.

www.ingramcontent.com/pod-product-compliance
Lightning Source LLC
Chambersburg PA
CBHW081107170526
45165CB00008B/2354